DEVELOPMENTS IN GEOTECHNICAL ENGINEERING 54

FOUNDATIONS ON EXPANSIVE SOILS

DEVELOPMENTS IN GEOTECHNICAL ENGINEERING 54

FOUNDATIONS ON EXPANSIVE SOILS

F.H. CHEN

Chen & Associates, 96 South Zuni, Denver, CO 80223, U.S.A.

ELSEVIER
Amsterdam — Oxford — New York — Tokyo   1988

ELSEVIER SCIENCE PUBLISHERS B.V.
Sara Burgerhartstraat 25
P.O. Box 211, 1000 AE Amsterdam, The Netherlands

Distributors for the United States and Canada:

ELSEVIER SCIENCE PUBLISHING COMPANY INC.
655, Avenue of the Americas
New York, NY 10010, U.S.A.

ISBN 0-444-43036-9 (Vol. 54)
ISBN 0-444-41662-5 (Series)

Printed in The Netherlands

To my wife Edna with love and appreciation.

Dr. Fu Hua Chen received a B.S.C.E. degree from the University of Michigan, a M.S.C.E. degree from the University of Illinois, and an Honorary Doctor of Science degree from Colorado State University. With both academic and pragmatic involvement, he has been a professor and guest lecturer at universities in China and in Colorado, and has more than fifty years experience in soil mechanics and foundation engineering. Dr. Chen was chief engineer of the Burma Road and the Koknor-Tibet Highway in China. Since he founded Chen & Associates, Inc., Consulting Soil Engineers, Denver, Colorado in 1961, the firm has completed more than 35,000 geotechnical projects and has grown to a staff of over 200 people in eight offices. The recipient of many honors and awards and a member or an officer in state and national professional societies, as well as an honorary member of the American Society of Civil Engineers. Dr. Chen is Chairman of the Board of Chen & Associates and continues to be active in the company's projects.

PREFACE

The problems associated with expansive soils are not widely appreciated outside the areas of their occurrence. The amount of damage caused by expansive soils is alarming. It has been estimated that the damage to buildings, roads, and other structures founded on expansive soils exceeds two billion dollars annually.

In the past 20 years, considerable progress has been made in understanding the nature of expansive soils. This new knowledge can be separated into two categories. The first emphasizes the theoretical approach and is the result of studies mostly by academic institutions. Such research involves soil mineralogy, structure, and modification. Academicians have also advanced new theories such as effective stress, soil suction, and osmotic pressure which reveal properties of swelling soils previously little known to engineers. The second category is concerned with the field performance of expansive soils with emphasis on design criteria and construction precautions for structures founded on expansive soil. Practical approaches of combating the swelling soils problem are mostly undertaken by soils engineers; therefore, they must offer practical and economical solutions to their clients so that the structure will be free from damaging foundation movement.

Unfortunately, present day knowledge of expansive soils has not reached a stage at which rational solutions can be assigned to the problem. It is difficult for the public to understand why the soils engineer is not capable of offering easy solutions. When the first crack appears in a structure, a lawsuit is threatened.

This book provides the practicing engineer with a summary of the state-of-the-art knowledge of expansive soils and practical solutions based upon the author's experience. Part I discusses theory and practice, and summarizes some of the theoretical physical properties of expansive soils. It also discusses various techniques employed to found structures on expansive soils such as drilled pier foundation, mat foundation, moisture control, soil replacement, and chemical stabilization. Part II presents typical case studies. The author has found that few records are available on the cause of structural distress, their remedial measures, and, more importantly, the degree of success after those measures have been completed.

In the last 15 years, the author has investigated many thousands of building sites in expansive soil areas in the Rocky Mountain region. He has also investigated over 1,000 cracked buildings and has suggested remedial

X

measures.   It is the author's hope that by sharing his knowledge and the
knowledge of other practicing engineers, a better understanding of expansive
soil problems can be achieved.

The author wishes to thank the entire staff of Chen and Associates for
sharing the work load while the author was devoting his time to writing this
book and also for the assistance given by them in the preparation of the
manuscript.   Many thanks to the various consulting firms, especially
Woodward-Clyde and Associates, Jorgensen and Hendrickson, Ketchum-Konkel-
Barrett-Nickel-Austin, and E.H. Tippets Company for allowing the publication
of their valuable findings.   Mr. Byron Eskesen has conducted most of the
field investigation and laboratory testing presented in this book.

Denver, Colorado
August, 1975

PREFACE

More than ten years have passed since the author completed his work on expansive soil problems. A great deal of advancement in this field has taken place since then. Most of the research has been concentrated in the area of heave prediction, identification of swelling soil and soil suction. Studies on unsaturated soil, finite element analysis and stress modeling also have commanded a great deal of attention. In the meantime, more and more nations are aware of the damage caused by expansive soils. It would be fair to say that more than half of the world is affected by swelling soils.

As a consulting geotechnical engineer, the author has seen little advancement in the practical application of theoretical approaches towards solving engineering problems. Both the geotechnical engineers and the structural engineers adhere to the initial concept that by isolating water, the problem can be resolved. Unfortunately, more and more structural failure has been reported. In the United States, the legal professions play an important role in the problems of expansive soils. It is the contention of the Court of Justice that someone should pay for the damage caused by expansive soils. The most logical victim is the geotechnical engineer.

In this revised edition, considerable new material has been added in each chapter, especially on highway pavement and soil suction. The author is indebted to Dr. John Nelson of Colorado State University and Mr. W.S. Lang for their contributions on soil suction, and to Roger Barker and Ken Criley of Chen & Associates for their contributions on geology and laboratory data.

For the preparation of the manuscript, the author is indebted to Beverly Noland and Carla Latuda.

Denver, Colorado
1988

CONTENTS

PART I
THEORY AND PRACTICE

CHAPTER 1 – NATURE OF EXPANSIVE SOILS

CHAPTER 2 – MECHANICS OF SWELLING

CHAPTER 4 DRILLED PIER FOUNDATION

CHAPTER 5 - FOOTING FOUNDATIONS

CHAPTER 8 - SOIL STABILIZATION

CHAPTER 9 - SOIL SUCTION

CHAPTER 10 - INVESTIGATION OF FOUNDATION MOVEMENT

        Drainage. 331

REMEDIAL CONSTRUCTION. 333

        Symptoms and causes 333
        Structural considerations 335
        Pier replacement. 337

LEGAL ASPECT 339

        Responsibility. 340
        Act of God. 341
        The pattern of the lawsuit. 342
        Future trend. 344

PART II
CASE STUDIES

CASE VI – AN ANATOMY OF A LAWSUIT

PART I

THEORY AND PRACTICE

Chapter 1

NATURE OF EXPANSIVE SOILS

INTRODUCTION

The problem of expansive soils was not recognized by soil engineers until the latter part of 1930. Prior to 1920, most of the lightly loaded buildings in the United States consisted of frame dwellings. Such structures could withstand considerable movement without exhibiting noticeable cracks. By 1930, brick veneer residences became widely used. It was then that the owner found cracks developing in the brick course. The damages were attributed to shoddy construction and settlement of the foundation at one corner, without recognition of the role of expansive soils.

The U.S. Bureau of Reclamation [1-1]* first recognized the swelling soil problem in 1938 in connection with a foundation for a steel siphon at its Owyhee Project in Oregon. Since that time, engineers realized the cause of damage was sometimes other than settlement. The increasingly extensive use of concrete slab-on-ground construction, after 1940, has further increased the damage to structures caused by expansive soils.

Today, there is a worldwide interest in expansive clays and shales. Engineers from Canada, Australia, South Africa, Israel, and the United States have contributed immensely to the knowledge and the proper design for structures on expansive soils. The first significant national conference on expansive clay probably was the one held at the Colorado School of Mines in Golden, Colorado in 1959. The International Research and Engineering Conference on Expansive Soils held its first and second conference at Texas A & M University in 1965. Since then, the International Conference on Expansive Soils has been held every four years.

ORIGIN OF EXPANSIVE SOILS

The origin of expansive soils is related to a complex combination of conditions and processes that result in the formation of clay minerals

*Numbers in brackets refer to items in the references at the end of each chapter.

having a particular chemical makeup which, when in contact with water, will
expand. Variations in the conditions and processes may also form other clay
minerals, most of which are nonexpansive. The conditions or processes which
determine the clay mineralogy include composition of the parent material and
degree of physical and chemical weathering to which the materials are
subjected.

Parent material

The constituents of the parent material during the early and
intermediate stages of the weathering process determine the type of clay
formed. The nature of the parent material is much more important during
these stages than after intense weathering for long periods of time [1-2].

G.W. Donaldson [1-3] classified the parent materials that can be
associated with expansive soils into two groups. The first group comprises
the basic igneous rocks and the second group comprises the sedimentary rocks
which contain montmorillonite as a constituent.

The basic igneous rocks are comparatively low in silica, generally
about 45% to 52%. Rocks which are rich in metallic bases such as the
pyroxenes, amphiboles, biotite and olivine fall within this category. Such
rocks include the gabbros, basalts and volcanic glass. Geographically, the
rocks of this type which have been associated with expansive soils are the
basalts of the Deccan Plateau in India; the dolerite sills and dikes; the
Bushveld niorite complex; the Stormberg basalt in South Africa; and the
niorites north of Pretoria and in the Great Dike of Southern Rhodesia.

The sedimentary rocks that contain montmorillonite as a constituent
include shales and claystones. Limestones and marls rich in magnesium can
also weather to clay. These constituents of the shales and claystones
contain varying amounts of volcanic ash and glass which were subsequently
weathered to montmorillonite. The volcanic eruptions sent up clouds of ash,
which fell on the continents and seas. Some of the fine-grained sediments
which accumulated to form these rocks undoubtedly contain montmorillonite
derived from weathering of continental igneous rocks and from ash which fell
on the continental areas. Geographically, the more noted rocks in this
category include the Pierre Shale and its counterparts, the Mancos and Bear
Paw Shales and the Denver and Laramie Formations in North America. In
Israel, there are marls and limestones, and in South Africa the shales of
the Ecca series. Lacustrine and deltaic deposits of lesser distribution and

notoriety occur worldwide.   The Winnipeg clay of Canada and the Yazoo and
Beaumont clays of the Texas and Mississippi area are examples.

Weathering

The weathering processes by which clays are formed include physical,
biological and chemical processes.

Krumbein and Sloss [1-4] indicate that physical weathering changes the
particle size and bulk volume of the parent material with no significant
change in composition.   Chemical weathering causes a complete change in
physical and chemical properties, accompanied by an increase in bulk volume
caused both by the lesser density of the new compounds and by additional
porosity of the weathered aggregate.   Biological weathering is similar to
chemical weathering in that changes in both the state of the aggregate and
in chemical composition occur.

Physical weathering processes include expansion due to unloading,
crystal growth, thermal expansion and contraction, organic activity and
colloidal plucking.

Chemical processes of weathering include hydration, hydrolysis,
oxidation, carbonation and solution.   All of these processes depend on the
presence of water as the means by which they occur.

Since water is the prime source of chemical weathering, climatic
conditions play an important role in the rate and extent of weathering of
the parent material that can occur.   The climate affects the amount of water
available to infiltrate the subsurface materials and the temperature, which
in turn affects the vegetative growth.   Climate ultimately becomes a more
dominant factor in residual soil formation than does parent material.
[1-2].

Twenhofel [1-5] states that the formation of expansive clays or the
montmorillonite clays is favored by an alkaline environment and the absence
of leaching, the presence of ferromagnesium minerals in parent materials and
the presence of bases.   Prolonged leaching under high temperatures or
tropical conditions with ferric iron parent rocks favors formation of
minerals of the kaolinite group which are nonexpansive.   The presence of
potash in the parent material under these conditions results in the
formation of illite.

DISTRIBUTION OF EXPANSIVE SOILS

G.W. Donaldson [1-3] summarized the distribution of reported instances of expansive soils around the world in 1969.   The countries in which expansive soils have been reported are as follows:

Argentina	Iran
Australia	Mexico
Burma	Morocco
Canada	Rhodesia
Cuba	South Africa
Ethiopia	Spain
Ghana	Turkey
India	U.S.A.
Israel	Venezuela

Since that time, many other nations have reported significant findings of expansive soils.   Among them, the most prominent is the extensive research of expansive soils in China and the black cotton soil in Sudan as well as that of Cyprus, Jordan and Saudi Arabia.

Figure 1-1 indicates that the potentially expansive soils are confined to the semi-arid regions of the tropical and temperate climate zones. Expansive soils are in abundance where the annual evapotranspiration exceeds the precipitation.   This follows the theory that in semi-arid zones, the lack of leaching has aided the formation of montmorillonite.

Potentially expansive soils can be found almost anywhere in the world.   In the underdeveloped nations, many of the expansive soil problems may not have been recognized.   It is to be expected that more expansive soil regions will be discovered each year as the amount of construction increases.

World problem of expansive soils

The problem of expansive soil is widespread throughout the five continents.   The following are the typical findings in each country:

Australia: The predominant soils in Australia are the podzolic soils, which do not normally show a great deal of expansion and shrinkage, but Australia does contain great areas of expansive soils; some of these areas are arid.

Gordon Aitchison [1-6] stated that of all the major cities in Australia, only Adelaide has severe problems.   Even though the damage caused

Figure 1-1 Distribution of reported instances of heaving

by expansive soils is moderate in a city of some 600,000 inhabitants, the aggregate damage associated with foundation cracks is a substantial amount.

Australia is by far the driest continent. It has a large area of desert, few major rivers and not more than 10 percent of the total water falling on the continent goes back to the sea as runoff.

Canada: The most troublesome expansive soils for shallow foundations on the Canadian prairies are those found in many deep lacustrine deposits from the Pleistocene age [1-7]. The Cretaceous shales from which expansive clay minerals were largely derived are usually deeply buried below glacial deposits and outcrops only in the deeper valleys. While they do pose stability problems on dams, canals and highways, these expansive clay shales rarely are of concern in the design or construction of shallow foundations.

Glacial lakes, such as Agassiz, left very deep deposits of highly expansive clay which was subsequently overlain with thin flood plain deposits in the vicinity of its current rivers. The Assiniboine River flows in oversized valleys that were once glacial spillways deeply cut into the Cretaceous deposits of Eastern Saskatchewan and Western Manitoba.

The wide range of climate and geology in Canada produces a great variety of foundation problems. In Western Canada, including Saskatchewan and Alberta, expansive clay problems are strongly evident. The soils in this region are generally desiccated. In addition, shallow basements placed on shallow footings are commonly used in this area of Canada. There have been very many cases where pressures of the expansive clays have caused lateral deflections of basement walls. Basement floors have been known to heave as much as 6 inches in 18 months [1-8].

The Eston clay of the southwestern part of Saskatchewan with a liquid limit of 94% and an average plasticity index of 63% indicates a high swelling potential. The swelling pressure varies from 10,000 to 30,000 psf. The Regina clay is a highly plastic lacustrine clay which is typically unsaturated, expansive and desiccated near the surface by evapotranspiration.

Shallow foundations in the City of Winnipeg, which is located where the Assiniboine River joins the Red River, have had a long history of heaving problems.

China: In the vast area in China, with complicated geological and topographical features, it is difficult to outline the expansive soil areas.

With few exceptions, the most frequently reported cases of severe expansive soil problems appear to be along longitudes of 22 to 28 degrees [1-9]. This zone covers the provinces of Yunna, Kweichow, Kwangsi and part of Szechuan as shown on Figure 1-2. In the north and northwest, the great loess deposit presents no swelling problems. In South China, the soft clays and the tropical red laterite also appear to have more settling problems rather than expansion problems. North of longitude 28 degrees in the Provinces of Hopei, Honan, Shangtung, Anhwei and Shensi, many cases of swelling soils have been reported. However, they appeared to be mild compared to those of Yunna and Kwangsi.

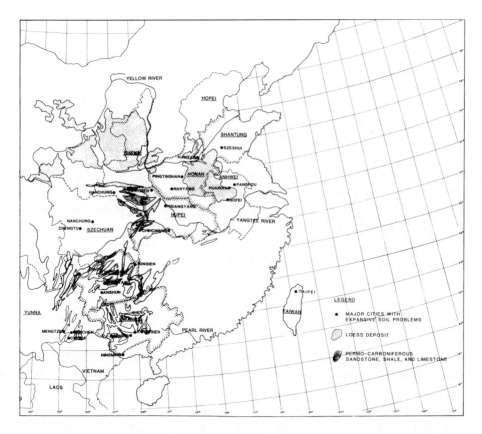

Figure 1-2 Distribution of expansive clays in China

The geological age of expansive soils in China generally falls within the category of Late Tertiary to Holocene. The formations generally were

derived from glacial, alluvial, residual and colluvial deposits.   [1-10]
The most severe swelling cases were reported at Mengtze, Yunna, a deposit
which originated from mudstone and claystone.   [1-11] Mentze borders Viet
Nam and is in a semi-tropical area.   Nanning, Kwangsi also has notorious
swelling soils. The rock formations and swelling characteristics in these
areas reflect the swelling of claystone shale in the Rocky Mountain states
in North America.

In contrast to the shale deposit in semi-tropical South China, swelling
soils found in North Szechuan and Anhwei are in relatively plain areas with
glacier or alluvial deposits.   The well-known Chengtu clay has long been
recognized for its swelling tendency.

The mineral composition of expansive clays in China consists mainly of
montmorillonite and illite.   At Mentze, Yunna; Nangning, Kwangsi; Hantan,
Hopei,   and   Pingtinshan,   Honan   montmorillonite   is   the   predominant
composition.   [1-12] As expected, more severe expansion problems have taken
place in these locations.   At Hofei, Anhwei; Chengtu, Szechuan; Yunhsien,
Hupei and Szeshui, Shangtang, illite is the predominant composition.   Only
mild expansion has taken place in these areas.

Israel: Professor Zeitlen [1-6] described Israel's climate as typically
Mediterranean, with a rainy season and a hot, dry summer one.   In the
northern portion of the country, the annual precipitation might vary between
2-4 inches but it might be as little as 0.8 inches per year in the south.

The soils are primarily alluvium or reworked transported alluvium which
originates from the weathering of either basalt or limestone.   In the clay
soils area, montmorillonite may be present in quantities ranging from 40 to
80 percent of the soil.   The plasticity index may vary from about 25-60 with
shrinkage limit as low as 7 having been recorded.   The clay soils are
usually desiccated and slickensided.

The most obvious difficulties in connection with expansive clays occur
with building foundations.   Both shrinkage and swelling occur.   Light
structures - even when founded on footings or piles penetrating through the
active zone - have been badly cracked.   In addition, movements have occurred
horizontally as well as vertically; piles have been completely sheared off
by these movements, and water and sewage lines have been broken to further
complicate the problems.   Also, there have been many cases where fills
beneath floors have heaved and damaged both the floors and the grade
beams. The problems, however, are not confined to light structures.

Swelling pressures have been enough to damage some large, rigid structures with fairly heavy foundations and piles.

Jordan: The climate of the area is of the Mediterranean, with cool, wet winters and hot, dry summers. The clay at depth is usually moist, but the topmost 6 to 10 feet becomes desiccated in the summer months. The expansive soils are derived mainly from beds of marl, limestone and chert in the underlying Cretaceous bedrock.

Highly expansive soils are reported in Northern Jordan, especially in Shameisani, [1-13]. Cracking of the internal and external walls ranged from hairline to 1.2 inches wide.

Saudi Arabia: Few are aware of the existence of expansive clays in Saudi Arabia where sandy deserts are the known geological feature. The behavior of the swelling soils in five regions is shown on Figure 1-3. [1-14]

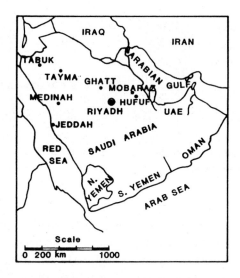

Figure 1-3 Some of the regions of swelling in Saudi Arabia

The regions of expansive soils in Saudi Arabia are located mainly at the northwestern portion of the Kingdom. Montmorillonitic clay with high swelling potential is found in Medinah region. Fissured clayey silt (shale) is found in Tabuk and Tayma region. The area is surrounded by the rock of Al Tubayq which belongs to the Ordovician, Silurian and Devonian Tabuk

formation. Hard gypsiferous clayey silt (laminated shale) is found in the
Al-Ghatt region. Calcareous clays of high salt content is typical of that
of the Hofuf region. The calcareous clay layer is underlain by a cavernous
sandy to marly limestone.

India: In India, large tracts are covered by expansive soils known as black
cotton soils, referred to as "Regur." [1-15] The major area of their
occurrence is the south Vindhyachal range covering almost the entire Duccan
Plateau. These soils cover an area of about 200,000 square miles and thus
form about 20% of the total area of India, as shown in Figure 1-4.

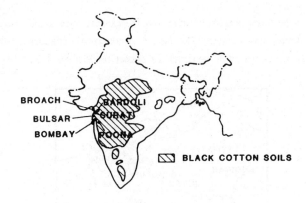

Figure 1-4 Distribution of black cotton soils in India. (After Katti)

    Krishan [1-16] reported that the black cotton soils in India exist as a
clayey to loamy soil, generally black and contains high amounts of alumina,
lime and magnesia with a variable amount of potash, low nitrogen and
phosphorus.   Its swelling property is due to the high content of
montmorillonite and beidellite groups of clay minerals.
    Singh [1-17] reported that the black cotton soils in India have a
predominance of montmorillonite clay mineral and are believed to have been
formed as a result of weathering of basalt. The soil profile typically
averages about 3 feet in thickness and in some regions reaches up to 25
feet.
    Geologically, the formation of black cotton soils is usually associated
with basalts.   However, with their occurrence on granitegneiss, shales,
sandstone and slates, limestone is also present. The range of liquid limit

for black cotton soils is 40 to 100, the plasticity index 20 to 60, and the shrinkage limit 9 to 14. Volumetric shrinkage on wet basis is reported to be about 40% to 50%.

Underreamed pile foundations have been successfully used in India for more than two decades and have been standardized for general use by the National Building Code of India in 1970.

South Africa: In South Africa, the problem of expansive soils was brought to the attention of the engineers as early as 1950. The South African Institution of Civil Engineers published the first symposium on expansive clays in 1957. Severe foundation movement problems were recorded at Leeuhof, Vereeniging and Pretoria in Transvaal, where the fluvio-lacustrine deposits are the source of swelling soils. The Ecca shale, covering a large part of South Africa, is responsible for the foundation movement problems at Odendaalsrus in the Orange Free States' goldfields.

The most highly expansive soils in South Africa are the black and grey subtropical clays developed as residual soils on mafic rocks. Also "heavy" expansive clays are developed on the "greenstone" i.e., the amphiboles, serpentine and phyllitic rocks. [1-18]

Sudan: Expansive soils cover an area approximately equal to one-third of Sudan's 2.6 million square kilometers. [1-19] The eastern area of Sudan, called the clay plain, consists of alkaline soils containing the clay mineral montmorillonite. The northern clay plain soils, locally termed black cotton soils, are weathered sediment derived from the volcanic Ethiopian highlands. The southern clay plain soils are similar but are considered residual soils. These clay soils, usually firm during dry months, become slick and impassable when rain occurs.

Figure 1-5 shows the location of the cities where expansive soils are known to have caused structural damage. Intensive damage has occurred in Khartoum. The clay plain indicated by the shaded area in Figure 1-5 had soils with 40% montmorillonite. The Atterberg limits are fairly uniform across the clay plain. The plasticity index with a mean of 45 indicates a medium-to-very high swell. The swellings under the surcharge load of 144, 500, 1000 and 9600 psf were found to be 8.2%, 4.32%, 3.83% and 0.11% respectively. The swelling pressure was found to be constant at the different surcharge pressures and a magnitude of 10 ksf. [1-20]

Black cotton soils occupy most of the Gezara Province, the Blue Nile Province and some areas of Kasalla Province, particularly the Gedaref

12   FOUNDATIONS ON EXPANSIVE SOILS

area.   The Ethiopian black soils are the continuation of Gedaref soils in
the eastern part of Sudan.   Such residual soils are derived from weathering
of basalt.

Ethiopia: Morin and Parry [1-21] reported about the Ethiopian black soils
which are the continuation of the Gedaref area soils in the eastern part of

Figure 1-5 Areas reporting expansive soil or damage in Sudan

the Sudan.   The Ethiopian black soils are residual soils, derived from the
weathering of basic volcanic rocks and are invariably clays or silty clays.
     Natural deposits of black cotton soils in the field are characterized
by a general pattern of cracks during the dry season of the year.   Cracks
about 2.5 inches wide and over 3 feet deep are not uncommon.   In deep
deposits, the cracks extend up to 10 feet or more.   During wet seasons, the
soil first expands horizontally, filling up the shrinkage cracks and thus
utilizing about two-thirds of the volumetric expansion.   The remaining
volumetric expansion causes vertical heaving of soil which may cause damage
to an overlying structure (Singh, 1967) [1-17].

Rhodesia: The climate is sub-tropical and is relatively uniform over the whole of the country with rainfall continued to the summer months.   About two-thirds of Rhodesia is covered by areas of granite and gneiss with frequent intrusions of dolerite and quartz.   The remainder of the country consists predominantly of sedimentary and metamorphic rocks with occasional intrusions of basalt.

C.P. Van der Merwe and M. Ahronovitz [1-22] used the cation exchange capacity known as E/C value for estimating the expansiveness of a soil.   A high E/C value with a high clay content provides a high degree of potential expansiveness.   Such soils known as the vertisol group are abundant in southern Rhodesia (Triangle Area) and western Rhodesia (Bulaways Area).

Britain: Britain's mild, damp climate is very different from those of countries in the humid tropics and the semi-arid zones where most expansive soils occur.   The clays, well known for giving foundation problems in London, Reading, Gault, Kimmeridge and Oxford, consist principally of interstratified micasmectite minerals, whereas the other soils consist largely of illite [1-23].

The shrinkage of clay soils following especially dry summers has caused damage to lightweight masonry buildings on shallow foundations.   Problems associated with expansive soils in Britain are largely caused by tree roots extracting soil moisture.

Spain: In Spain, many clay formations of sedimentary origin with high plasticity can be found.   In most parts of the country, the climate is arid and the evapotranspiration is several times greater than the precipitation, resulting in swelling phenomena.   Among the various regions where such phenomena have been observed, there are two provinces which may be regarded as typical: Andalucia and Madrid.   In the province of Madrid, the soils for the most part consist of montmorillonitic clays.   These soils reach a liquid limit of 250, though generally they do not go over 80.   In a great number of the metropolitan areas, the highly plastic clays are covered with a sufficient depth of sandy clay sediments, and, therefore, present no swelling problem.

Peru: Expansive soils in Peru have been attributed to the alteration of the arid, dry climate with cycles of heavy rainfall in those areas near the equator where it is very dry due to the lack of rainfall.   In the North Peru Region in areas more or less extended especially in Piura and Lambayeqye and in lesser degree in La Libertad, most of the expansive clays range from medium plastic to very plastic with liquid limits greater than 40 and plasticity indices higher than 20. In the case of Talara and Piura clay, the liquid limit reached 148 and plasticity reached 93. When such soils were tested after 24 hours from the beginning of saturation, it practically exploded, producing a very important, difficult-to-measure swelling. In most of the cases, the soils have high density that reached 114 pcf (2 ton/m^3). [1-24]

Venezuela: The first documented report of swelling clays in Venezuela came from the vicinity of the City of Coro where many buildings are badly cracked. [1-25]. In one instance near the city, shales with expansive properties were found. Some of these soils have swelling pressures of 13 tons per sq. ft. and occasionally up to 28 tons per sq. ft.

Mexico: Mexico City has a world-renowned reputation for settlement problems. The problem of expansive clays in Mexico is not considered very serious to date. So far, they have been encountered in only about five towns of rather medium size. Because of the nature of the construction, the problem is potentially more serious with new towns which are being constructed and small towns which are being expanded.

Distribution of expansive soils in the United States

In the United States, from the Gulf of Mexico to the Canadian Border and from Nebraska to the Pacific Coast, the abundance of montmorillonite is common in both clays and claystone shales.

The reported problem locations are mostly in the regionally abundant montmorillonite areas indicated in Figure 1-6. Many maps were presented on the distribution of expansive soils in the United States, including those of the Federal Highway Administration and the Corps of Engineers. The author questions that the maps depict the actual degree of severeness of expansive soils. The states that experience various degrees of expansive soil problems are listed as follows:

Severe: Colorado

                 Texas

                 Wyoming

Moderate: California

Utah

Nebraska

South Dakota

Mississippi

Mild: Oregon

Montana

Arizona

Oklahoma

Kansas

Alabama

The major concern with expansive soils exists generally in the western United States and the Gulf States.   In the northern and central United States, the expansive soil problems are primarily related to highly overconsolidated shales.   This includes the Dakotas, Montana, Wyoming and Colorado.

In Minneapolis, some expansive soil problems exist in the Cretaceous deposits along the Mississippi River and shrinkage/swelling problems exist in the lacustrine deposits in the Great Lakes area.   In general, however, expansive soils are not encountered regularly in the eastern parts of the central United States.

Figure 1-6.   General abundance of montmorillonite in near outcrop
bedrock formations.   (Modified from Tourtelot, 1973)

In eastern Oklahoma, Texas and the Gulf States, the problems encompass both shrinking and swelling. In these areas, moisture fluctuations over the different seasons cause shrinking during drier periods and swelling when they are wetted again. Volume change can take place over relatively short periods of time ranging from a few months to one or two years.

In central and western Texas and western Oklahoma, the problems include both overconsolidated shales as well as some shrinking and swelling problems.

In the Los Angeles area, the problem is primarily one of desiccated alluvial and colluvial soils. Many of the soils have a volcanic origin and have been highly desiccated in the natural state.

Austin and San Antonio, Texas: Most of the Austin area is underlain by Cretaceous sediments, ranging from the Taylor Formation in the eastern part of Austin to the Glen Rose Formation west of Austin. [1-26] The deposits that cause the foundation problems in the Austin area are the Taylor, Eagle Ford, Pepper and Del Rio. The clays generally have liquid limits ranging from 50% to 75% and a plasticity index ranging from 30% to 60%. Swelling pressure generally ranges from 2,000 psf to 8,000 psf, but some values over 10,000 psf have been measured.

San Antonio's geologic environment is quite similar to that of Austin. The principal foundation soils range from Taylor and Austin Formations in the north to more recent Pleistocene deposits in the southern part of the city. The Taylor Formation causes most of San Antonio's foundation problems.

Most of the residences in the San Antonio area were constructed on reinforced concrete slabs. The slabs are 4 inches thick supported by reinforced concrete beams. For residences on the Taylor Formation, 70% reported none-to-light damage and 30% reported important damage.

Jackson, Mississippi: Many cases of structural damage have been observed in the vicinity of Jackson that range from slight movements to displacements of as much as one foot in the underlying Yazoo clay formation. [1-27]

Jackson Prairie runs in a generally east-west direction across the state of Mississippi. The predominant soil of this belt is a flat, stiff marine clay of the Jackson group called Yazoo clay because of its outcropping along the Yazoo River Basin. The Yazoo clay formation was deposited in glacial times with a thickness of about 400 feet in the Jackson area.

The upper 10 to 15 feet of the Yazoo clay was weathered with shrinkage cracks. The cracks were closed when the overburden was placed but remained as joints or slickensides. At greater depths, the Yazoo clay is unweathered and not jointed.

Atterberg limits of Yazoo clay indicate a liquid limit from 50 to 120 psf and plasticity indices from 30 to 80 psf with swelling pressure from 1,500 to 3,500 psf. When wetted and air-dried, Yazoo clay slakes quickly, while the structure of the material previously kept at a natural water content is affected very little by the addition of water.

Hawaii: The Hawaiian Islands are formed from a chain of mountaintops of old volcanoes in the middle of the Pacific. Since the islands were built up by a series of thinly bedded flows, a cut through the lava formation may reveal alternating layers of soil and lava rock, each layer being generally less than 5 to 10 feet thick. [1-28]

Lava formations vary considerably. The southern half of the island of Oahu is spotted with secondary ash cones. Montmorillonite clays are usually associated with ash deposits; so whenever ash cones occur, the surface soils within the cone and at the toe of the cone are usually expansive. Expansive soils can be found at and around Diamond Head Crater, Kokohead, Allamann, etc.

Typical characteristics of Hawaii clay are a high moisture content (over 50%), low dry density (60 to 80 pcf) and very high index properties (liquid limit 150 to 200%). Such soils in their natural state possess very low swell potential. It is only after extensive drying, reducing its natural moisture content to half and increasing the dry density to almost 100%, that subsequent wetting will cause the soil to swell considerably.

Denver, Colorado: At the end of the Cretaceous time, the uplift in Colorado that built the Rocky Mountains began as part of the Laramide Orogeny [1-29]. As the mountains rose, the land east of the Front Range subsided, forming the Denver Basin. The Denver Basin was a site of deposition for sediments eroded from mountains including the Arapahoe, Denver, and Dawson Formations. The Denver Formation is highly variable in texture and composition, consisting mainly of light gray to brown tuffaceous silty claystone. The Dawson Formation is similar to the Denver but sandier and contains less volcanic materials. The weathered volcanic material in the Denver Formation commonly swells when wetted and is, thus, the cause of major engineering problems in the Denver area.

In some parts of the Denver area, bedrock appears at the surface and is covered by thin colluvium and residuum formed by in situ weathering. The depth of overburden varies from less than a foot to over 100 feet. The elevation of the eroded bedrock can change dramatically over a short distance.

Moderately swelling soils are estimated to be present in the surficial materials over about 50% of the Denver area, particularly in the south, southeast and western portions. Approximately 25% of the area is affected by a high-to-very high swell potential. Swelling soils typically have liquid limits of 45% to 65% and a plasticity index of 25% to 35%. The soils swell 3% to over 10% under a normal load of 1,000 psf, with swelling pressures reached as high as 50,000 psf.

DAMAGE CAUSED BY EXPANSIVE SOILS

Jones and Holtz reported in ASCE in 1973 [1-30] the estimated damage attributed to expansive soil movement as follows:

Construction category	Estimated average annual loss, millions of dollars
Single-family homes	$ 300
Commercial buildings	360
Multi-story buildings	80
Walks, drives, parking areas	110
Highways and streets	1,140
Underground utilities and service	100
Airports	40
Urban landslides	25
Others	100
Total	$ 2,255

The National Science Foundation in 1978 [1-31] in a study on "Building Losses from Natural Hazards" listed six major natural hazards. These are earthquake, landslide, expansive soils, hurricane, tornado and flood. The study pointed out that expansive soils tie with hurricane wind/storm surge for second place among America's most destructive natural hazards in terms of dollar losses to buildings.   Its destructive impact is currently surpassed only by that of riverine flood.

The study pointed out that few people have ever heard of expansive soils. Even fewer realize the magnitude of the damage they cause. In most cases, it takes a professional soil engineer to confirm the existence of such damage and evaluate its probable behavior.

Over one-fifth of the nation's families live on such soil, and no state is free from a significant amount of it. According to the study, it is projected that by the year 2,000, losses due to expansive soils can exceed 4.5 billion annually as shown on Figure 1-7.

Analysis of the damage of structures is extremely difficult to evaluate. It takes both the efforts of a geotechnical and a structural engineer to research its causes. Generally, the following constitute the different kinds of damage:

1.   Damage unrelated to foundation movement

2.   Damage caused by error of foundation design.

3.   Damage caused by construction defect or lack of inspection.

4.   Damage which cannot be avoided or explained with present-day knowledge on expansive soils.

In the above listed items, damage caused by structural movement has been discussed in detail in chapter 10.   Damage caused by design and construction will be fully discussed under separate headings such as drilled piers, slab foundations, etc. The author would like to stress that with present-day knowledge on expansive soils, it is not always possible to design and construct a structure within a practical economic scope to ensure that there will be no foundation movement.

As far the as cracking of structures is concerned, the evaluation of its seriousness depends on the owner's discretion. Some owners will be greatly disturbed by a few hairline cracks and will diligently seek legal advice. Others will not be bothered by minor foundation movements and take for granted that such is expected in an expansive soil area. The degree of concern also varies with the occurrence place. In underdeveloped nations, building distresses are seldom reported. In highway construction, pavement movement and cracking in secondary roads are usually left to maintenance and seldom reported. At the same time, airport administration sets strict requirements on the smoothness of runways, with which it is difficult to comply in an expansive soil area.

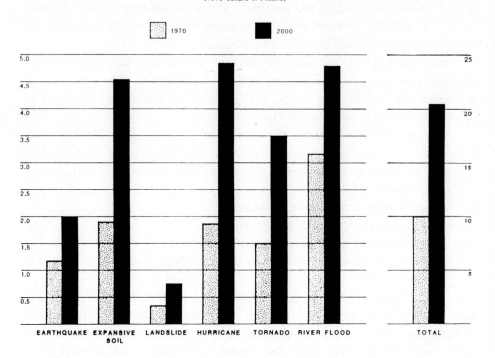

Figure 1-7 Average total annual building losses under 1970 and year
2000 conditions, (1978 dollars in billions).

CLAY MINERALS

Most soil classification systems arbitrarily define clay particles as
having an effective diameter of two microns (0.002 mm) or less. Particle
size alone does not determine clay mineral. Probably the most important
grain property of fine-grained soils is the mineralogical composition
[1-32]. For small size particles, the electrical forces acting on the
surface of the particles are much greater than the gravitational force.
These particles are in the colloidal state. The colloidal particle consists
primarily of clay minerals that were derived from parent rock by weathering.

The three most important groups of clay minerals are montmorillonite,
illite, and kaolinite, which are crystalline hydrous aluminosilicates.
Montmorillonite is the clay mineral that presents most of the expansive soil
problems.

The name "montmorillonite" is used currently both as a group name for all clay minerals with an expanding lattice, except vermiculite, and also as a specified mineral name [1-33].

Absorption of water by clays leads to expansion. From the mineralogical standpoint, the magnitude of expansion depends upon the kind and amount of clay minerals present, their exchangeable ions, the electrolyte content of the aqueous phase, and the internal structure.

Formation of clay minerals

The clay minerals are formed through a complicated process from an assortment of parent materials. The parent materials include feldspars, micas, and limestone. The alteration process that takes place on land is referred to as weathering and that on the sea floor or lake bottom as halmyrolysis. The alteration process includes disintegration, oxidation, hydration, and leaching.

Tourtelot [1-34] pointed out that the setting for the formation of montmorillonite is extreme disintegration, strong hydration, and restricted leaching. The situations in which montmorillonite can form require that leaching be restricted, so that magnesium, calcium, sodium, and iron cations may accumulate in the system. Thus, the formation of montmorillonitic minerals is aided by an alkaline environment, presence of magnesium ions, and a lack of leaching. Such conditions are favorable in semi-arid regions with relatively low rainfall or highly seasonal moderate rainfall, particularly where evaporation exceeds precipitation. Under these conditions, enough water is available for the alteration process, but the accumulated cations will not be removed by flush rain.

The parent minerals for the formation of montmorillonite often consist of ferromagnesium minerals, calcic feldspars, volcanic glass, and many volcanic rocks. Bentonite is a clay composed primarily of montmorillonite which has been formed by the chemical weathering of volcanic ash. Swelling clays are commonly referred to as bentonitic soils by laymen. Since commercial bentonite is white, the white calcium streaks present in stiff clays are often mistaken for bentonite. Actually, clays with an abundance of calcium seldom exhibit swelling characteristics.

Cation exchange

Clay minerals have the property of sorbing certain anions and cations and retaining them in an exchangeable state. The exchangeable ions are held around the outside of the silica-alumina clay-mineral structural unit, and the exchange reaction does not affect the structure of the silica-alumina pocket. In clay minerals, the most common exchangeable cations are Ca^{++}, Mg^{++}, H^+, K^+, NH_4^+, Na^+, frequently in about that order of general relative abundance.

The existence of such charges is indicated by the ability of clay to absorb ions from the solution. Cations (positive ions) are more readily absorbed than anions (negative ions); hence, negative charges must be predominant on the clay surface. A cation, such as Na^+, is readily attracted from a salt solution and attached to a clay surface. However, the absorbed Na^+ ion is not permanently attached; it can be replaced by K^+ ions if the clay is placed in a solution of potassium chloride KCL. The process of replacement by excess cations is called cation exchange [1-35].

The cation exchange capacity is the charge or electrical attraction for cation per unit mass as measured in millequivalent per 100 grams of soil.

The cation exchange capacity of different types of clay minerals may be measured by washing a sample of each with a solution of a salt such as ammonium chloride NH_4CL and the amount of adsorbed NH_4^+ by measuring the difference between the original and the final concentration of the washing solution.

Typical ranges of cation exchange capacities of various clay minerals are shown in Table 1-1.

From Table 1-1, it is seen that montmorillonites are 10 times as active in absorbing cations as kaolinites. This is caused by the large net negative charge carried by the montmorillonite particle and its greater specific surface as compared with kaolinite and illite.

Certain relationships exist between soil properties such as the Atterberg limits, the type of clay mineral, and the nature of the adsorbed ion. Table 1-2 indicates the liquid limit and the plasticity index of each group of clay minerals. From Tables 1-1 and 1-2, it is seen that the cation exchange capacity of a clay has a definite relationship with the Atterberg limits. The greater the cation exchange capacity of clay, the greater the effect of changing the adsorbed cation.

Cation exchange phenomenon takes place in everyday life. A simple and well known example of the ion exchange reaction is the softening of water by the use of permutites or carbon exchangers. The basic principle involved in the chemical stabilization of expansive soil is the increase in the ionic concentration in the free water and base exchange phenomenon.

Table 1-1. Ranges of cation exchange capacities of various clay minerals

	Kaolinite	Illite	Montmorillonite
Particle thickness	0.5 - 2 microns	0.003 - 0.1 microns	Less than 9.5 A
Particle diameter	0.5 - 4 microns	0.5 - 10 microns	0.05 - 10 microns
Specific surface (sq. meter/gram)	10 - 20	65 - 180	50 - 840
Cation exchange capacity (milliequivalents per 100g)	3 - 15	10 - 40	70 - 80

(After Woodward-Clyde & Associates, 1967)

Table 1-2. Atterberg limit values of clay minerals with various adsorbed cations

Cation	Na^+ Liquid limit, percent	Na^+ Plasticity index, percent	K^+ Liquid limit, percent	K^+ Plasticity index, percent	Ca^{++} Liquid limit, percent	Ca^{++} Plasticity index, percent	Mg^{++} Liquid limit, percent	Mg^{++} Plasticity index, percent
Clay mineral								
Kaolinite	29	1	35	7	34	8	39	11
Illite	61	27	81	38	90	50	83	44
Montmorillonite	344	251	161	104	166	101	158	99

(After W.A. White, 1958)

Clay structure

Philip Low [1-36] pointed out the two fundamental molecular structures as the basic units of the lattice structure. These are the silica tetrahedron and the alumina octahedron.

The silica tetrahedron consists of a silicon atom surrounded tetrahedrally by oxygen ions as shown on Figure 1-8a. The alumina octahedron consists of an aluminum atom surrounded octahedrally by six oxygen ions as shown on Figure 1-8b. When each oxygen atom is shared by two tetrahedra, a plate-shaped layer is formed. Similarly, when each aluminum atom is shared by two octahedron, a sheet is formed.

The silica sheets and the alumina sheets combine to form the basic structural units of the clay particle. Various clay minerals differ in the stacking configuration.

The results of studies using the electron microscope and X-ray diffraction techniques show that the clay minerals have a lattice structure in which the atoms are arranged in several sheets similar to the pages of a book. The arrangement and the chemical composition of these sheets determine the type of clay mineral. The basic building blocks of the clay minerals are the silica tetrahedron and the alumina octahedron. The blocks combine into tetrahedral and octahedral sheets to produce various types of clays.

Kaolinite is a typical two-layer mineral having a single tetrahedral sheet joined by a single octahedral sheet to form what is called a 2 to 1 lattice structure.

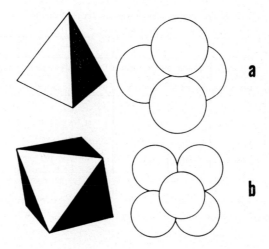

Figure 1-8. Polyhedra composing the structure of montmorillonite: (a) the silica tetrahedron, (b) the alumina octahedron. (After Philip Low, 1973).

Montmorillonite is a three-layer mineral having a single octahedral sheet sandwiched between two tetrahedral sheets to give a 2 to 1 lattice structure as shown on Figure 1-9.

Illite has a similar structure with that of montmorillonite, but some of the silican atoms are replaced by aluminum, and, in addition, potassium ions are present between the tetrahedral sheet and adjacent crystals.

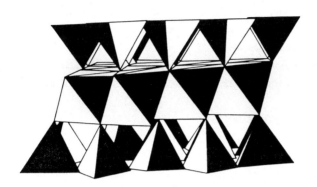

Figure 1-9. Model of a layer of montmorillonite. (After Philip Low, 1973)

In the clay-water air system, the water within the clay is called adsorbed water; the water and ions with the clay lattice constitute the diffuse double layer. Two forces exist in the system, the attractive and the repulsive forces.

The closer the dipolar water molecules and cations are to the flat plate surface, the more strongly they are attracted. At small interlayer distances, two attractive forces predominate.

1.   Electrostatic force - depends on the composition of the mineral.

2.   Van der Waals' force - depends on the distance between the layers.

The high concentration of cations near the surface of the clay particle creates a repulsive force between the diffuse double-layer system. The interlayer solution has a higher concentration of dissolved electrolyte than the external solution and the subsequent entry of water by osmosis. The resulting repulsive pressure is, therefore, the osmotic pressure.

The double-layer theory assumes that the clay particle is a flat, charged condenser plate and the ions are assumed to be non-interacting point charges. Hence, it is possible to use Poisson's equation from the theory of electrostatics. By combining Poisson's equation with Boltzmann's equation of osmotic pressure, the resulting equation is the Poisson-Boltzmann equation and is the basic differential equation of the double-layer theory. The typical result of the integration of the Poisson-Boltzmann

equation is given in Figure 1-10, in which the surface charge density is
determined by dividing the cation ion exchange capacity by the surface
area.   It is seen from Figure 1-10 that the calculated repulsive pressures
increase rapidly as the half-distance between the particles decreases.

Warkentine and Bolt [1-37] observed that experimental curves of
swelling pressure versus interlayer half-distance for Na-montmorillonite has
the same shape as Figure 1-10.

The basic relation between dry density and swelling pressure developed
by the author in Figure 2-22 also assumes the same pattern.

RECOGNITION OF EXPANSIVE SOILS

There are three different methods of classifying potentially expansive
soils.   The first, mineralogical identification, can be useful in the
evaluation of the material but is not sufficient in itself when dealing with
natural soils.   The various methods of mineralogical identification are
important in a research laboratory in exploring the basic properties of
clays, but are impractical and uneconomical for practicing engineers.

Another group includes the indirect methods, such as the index
property, PVC method, and activity method which are valuable tools in
evaluating the swelling property.   Soil suction may prove to be very useful
with more general application and improved testing techniques.   None of the
indirect methods should be used independently.   Erroneous conclusions can be
drawn without the benefit of direct tests.

The third method, direct measurement, offers the most useful data for a
practicing engineer.   The tests are simple to perform and do not require any
costly and exotic laboratory equipment.   A word of caution should be
introduced here.   Testing should be performed on a number of samples rather
than a few to avoid erroneous conclusions.

Mineralogical identification

The mineralogical composition of expansive soils has an important
bearing on the swelling potential as explained under "Clay Structure." The
negative electric charges on the surface of the clay minerals, the strength
of the interlayer bonding, and the cation exchange capacity all contribute
to the swelling potential of the clay.   Hence, it is claimed by the clay
mineralogist that the swelling potential of any clay can be evaluated by

identification of the constituent mineral of this clay. The five techniques
which may be used are as follows:

    X-ray diffraction,
    Differential thermal analysis,
    Dye adsorption,
    Chemical analysis, and
    Electron microscope resolution.

The various methods listed above should generally be used in
combination. Using combinations of the methods, the different types of clay
minerals present in a given soil can be evaluated quantitatively.
Unfortunately, though a great deal of research has been done in the various
fields of mineralogical study, the test results require expert
interpretation and the specialized apparatus required are costly and not
economically available in most soil testing laboratories. A brief
description of the various techniques is as follows:

X-Ray Diffraction Method. The X-ray diffraction method used in determining
the proportion of the various minerals present in a colloidal clay consists
essentially of comparing the ratios of the intensities of diffraction lines
from the different minerals with the intensities of lines from the standard
substance. G.W. Brindley [1-38] claimed that the use of self-recording
counter spectrometers in lieu of photographic techniques increases
considerably both the accuracy and the convenience of the X-ray method.
Brindley also believes that the X-ray method for quantitative determinations
should be applied with considerable circumspection, and that in favorable
cases the possibility of identifying species by X-ray analysis can be
regarded with restrained optimism.

Differential Thermal Analysis. Differential thermal analysis when used in
conjunction with X-ray diffraction and chemical analysis enables the
identification of otherwise difficult materials. It is well established as
a technique for the control of materials which undergo characteristic
changes on heating. The use of differential thermal analysis technique in
identifying expansive soil is not always accurate [1-39].

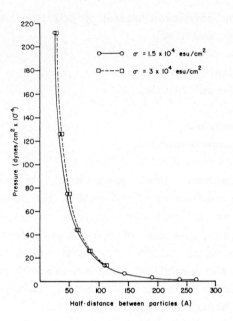

Figure 1-10. Calculated repulsive pressures at different half–distances
              between adjacent montmorillonite particles (layers) for
              two values of the surface charge density (s). (After
              Philip Low)

Dye Adsorption. Dyestuffs and other reagents which exhibit characteristic
colors when adsorbed by clay have been used to identify clay. When a clay
sample has been pretreated with acid, the color assumed by the adsorbed dye
depends on the base exchange capacity of the various clay minerals
present. The presence of montmorillonite can be detected if its amount is
greater than about 5 to 10 percent. The relatively simple testing procedure
and speed of dye staining tests compared with X-ray diffraction and
differential thermal analysis justify wider application of the color method.

Chemical Analysis. Chemical analysis can be a valuable supplement to other
methods such as X-ray analysis in identifying clays. In the montmorillonite
group of clay minerals, chemical analysis can be used to determine the
nature of isomorphism and to show the origin and location of the charge on
the lattice. According to Kelley [1-40], the isomorphous character of the
montmorillonite group can probably be shown in no other way. The

isomorphism involves three basic variations in the substitution: the
substitution for Al for Si in tetrahedral positions in the lattice; the
substitution of Fe for Al in the octahedral coordination; and the
substitution of Mg for Al in the octrahedral positions.

Electron Microscope Resolution.   Microscopic examination of clay minerals
offers a direct observation of the material. Two clays may give the same X-
ray pattern and the same differential thermal curve but will show distinct
morphological characteristics under electron microscope resolution.   The
main purpose of the microscopic examination is to determine mineralogic
composition, texture, and internal structure.

    Ravina [1-41] made extensive study of the mineralogical composition of
expansive clays by the use of the scanning electron microscope.   It showed
that the nonswelling clays appear as flat, relatively thick plates while
montmorillonites have a crinkly, ridged, honeycomb-like texture.   It might
be possible to evaluate some properties of the expansive soil by observing
the degree of crinkling and interparticle bonding from scanning an electron
microscope.

Single index method

    Simple soil property tests can be used for the evaluation of the
swelling potential of expansive soils.   Such tests are easy to perform and
should be included as routine tests in the investigation of building sites
in those areas having expansive soil.   Such tests may include:

    Atterberg limits tests,
    Linear shrinkage tests,
    Free swell tests, and
    Colloid content tests.

Atterberg Limits.   Holtz and Gibbs [1-42] demonstrated in 1956 that the
plasticity index and the liquid limit are useful indices for determining the
swelling characteristics of most clays. Seed, Woodward, and Lundgren [1-42]
have demonstrated that the plasticity index alone can be used as a
preliminary indication of the swelling characteristics of most clays.

    Since the liquid limit and the swelling of clays both depend on the
amount of water a clay tries to imbibe, it is not surprising that they are
related.

The relation between the swelling potential of clays and the plasticity index can be established as follows:

Swelling potential	Plasticity index
Low	0 - 15
Medium	10 - 35
High	20 - 55
Very high	35 and Above

While it may be true that high swelling soil will manifest high index property, the converse is not true.

Linear Shrinkage. The swell potential is presumed to be related to the opposite property of linear shrinkage measured in a very simple test. In theory it appears that the shrinkage characteristics of the clay should be a consistent and reliable index to the swelling potential.

Altmeyer suggested the following in 1955 [1-44] as a guide to the determination of potential expansiveness for various values of shrinkage limits and liner shrinkage:

Shrinkage limit as a percentage	Linear shrinkage as a percentage	Degree of expansion
Less than 10	Greater than 8	Critical
10 - 12	5 - 8	Marginal
Greater than 12	0 - 5	Non-critical

Recent research, however, failed to show conclusive evidence of the correlation between swelling potential and shrinkage limits.

Free Swell. Free swell tests consist of placing a known volume of dry soil in water and noting the swelled volume after the material settles, without any surcharge, to the bottom of a graduated cylinder. The difference between the final and initial volume, expressed as a percentage of initial volume, is the free swell value. The swell test is very crude and was used in the early days when refined testing methods were not available.

Experiments conducted by Holtz [1-42] indicated that a good grade of high-swelling commercial bentonite will have a free swell value of from 1200

to 2000 percent. Holtz suggested that soils having a free-swell value as
low as 100 percent can cause considerable damage to lightly loaded
structures, and soils having a free-swell value below 50 percent seldom
exhibit appreciable volume change even under very light loadings.

Colloid Content. The grain size characteristics of a clay appear to have a
bearing on its swelling potential, particularly the colloid content. Seed,
Woodward, and Lundgren [1-43] believed that there is no correlation between
swelling potential and percentage of clay sizes. For a given clay type,
however, the amount of swell will increase with the amount of clay present
in the soil as shown on Figure 1-11.

For any given clay type, the relationship between the swelling
potential and percentage of clay size can be expressed by the equation:

$$S = KC^X$$

where: S = Swelling potential, expressed as a percentage of swell
                  under 1 psi surcharge for a sample compacted at optimum
                  moisture content to maximum density in standard AASHO
                  compaction test,
              C = Percentage of clay sizes finer than 0.002 mm.
              X = An exponent depending on the type of clay, and
              K = Coefficient depending on the type of clay.

Where the quantity of the clay size particles is determined by a
hydrometer test, the quality or kind of colloid, which is reflected by X and
K in the above equation, controls the amount of swell. The colloid content
as well as the Atterberg limits should be included in the routine laboratory
investigation on expansive soils.

Classification method

By utilizing routine laboratory tests such as the Atterberg limits,
colloid contents, shrinkage limits, and others, the swelling potential can
be evaluated without resorting to direct measurement. Some of these methods
are as follows:

USBR Method - Developed by Holtz and Gibbs [1-42], this method is based on
the simultaneous consideration of several soil properties. The typical

relationships of these properties with swelling potential are shown on
Figure 1-12.

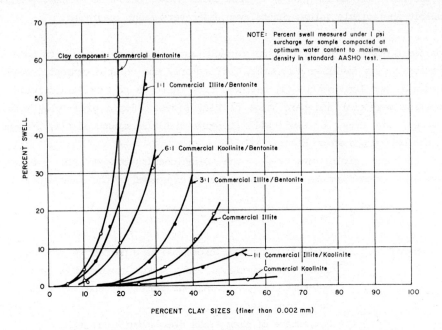

Figure 1-11. Relationship between percentage of swell and percentage of
             clay sizes for experimetal soils. (After Seed, Woodward &
             Lundgren).

Based on the curves presented in Figure 1-12, Holtz [1-45] proposed the
identification criteria of expansive clay as shown on Table 1-3.

Table 1-3. Data for making estimates of probable volume changes for expansive soils

Data from index tests*			Probable expansion, percent total vol. change	Degree of expansion
Colloid content, percent minus 0.001 mm	Plasticity index	Shrinkage limit		
>28	>35	<11	>30	Very high
20-13	25-41	7-12	20-30	High
13-23	15-28	10-16	10-30	Medium
>15	<18	>15	<10	Low

*Based on vertical loading of 1.0 psi. (After Holtz & Gibb)

It should be pointed out that Figure 1-12 is based on actual expansion tests for only 45 undisturbed and remolded samples and, therefore, the data accumulated is not sufficient to form an accurate empirical relationship between measured expansion and three indicator tests.   Especially,

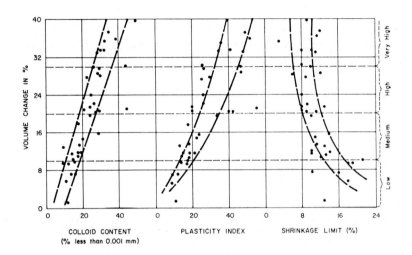

Figure 1-12.   Relation of volume change to colloid content, plasticity
               index, and shrinkage limit (air-dry to saturated condition
               under a load of 1 lb. per sq. in.)   (After Holtz and Gibbs).

consideration should be given to the differentiation of soil behavior between undisturbed and remolded samples.

The author has over the past 30 years performed many thousands of tests on potential swell and index properties.   From the test results of 321 undisturbed samples, a regression curve can be fitted as shown on Figure 1-13.   The relationship between the swell potential and the plasticity index can be expressed as follows:

$$S = Be^{A(PI)}$$

in which   A = 0.0838, and

           B = 0.2558

Figure 1-13 shows that with the increase of the plasticity index, the increase of swelling potential is much less than predicted by Holtz and

Gibbs or from Seed, Woodward and Lundgren. All tests refer to a surcharge
pressure of 1 psi with a moisture content between 15 and 20 percent and the
dry density between 100 and 110 pcf.

Activity Method.   The activity method proposed by Seed, Woodward, and
Lundgren [1-43] was based on remolded, artificially prepared soils composed
of twenty-three mixtures of bentonite, illite, kaolinite, and fine sand.
The expansion was measured as the percentage of swell on soaking from 100
percent maximum density and optimum moisture content in a standard AASHO
compaction test under a surcharge of 1 psi.   The activity for the
artificially prepared sample was defined as:

$$\text{Activity} = \frac{PI}{C-10}$$

In the above, C denotes the percentage of clay size finer than 0.002 mm.
The proposed classification chart is shown on Figure 1-14.
        The activity method appears to be an improvement over the USBR method
in that the shrinkage limit did not enter in the evaluation of swell
potential.   Also, an attempt has been made to differentiate between
undisturbed and remolded samples.

Indirect measurement

    Indirect measurement of swelling potential of expansive soils has been
approached by many investigations. The Ladd and Lambe method aided by a PVC
meter is probably the simplest and quickest method, while the soil suction
method is considered to be a new approach toward the measurement of swelling
potential and swelling pressure.

PVC Meter. The determination of the potential volume change (PVC) of soil
was developed by T.W. Lambe under the auspices of the Federal Housing
Administration [1-46]. Remolded samples were specified. The sample was
first compacted in a fixed ring consolidometer with a compaction effort of
55,000 ft.-lbs per cu. ft. Then an initial pressure of 200 psi was applied,
and water added to the sample which is partially restrained from vertical
expansion by a proving ring. The proving ring reading is taken at the end
of 2 hours. The reading is converted to pressure and is designated as the
swell index. From Figure 1-15, the swell index can be converted to

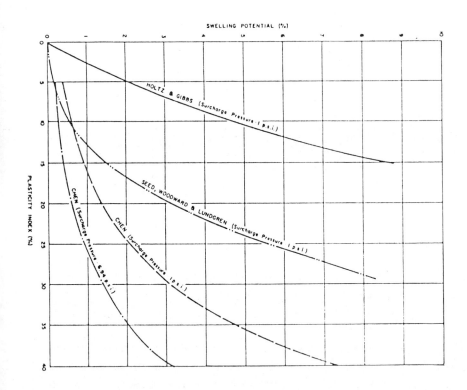

Figure 1-13. Relationship of volume change to plasticity index as
predicted by Holtz, Seed, and Chen.

potential volume change.   Lambe established the following categories of PVC
rating:

PVC Rating	Category
Less than 2	Non-critical
2 - 4	Marginal
4 - 6	Critical
Greater than 6	Very critical

The PVC meter method has been widely utilized by the Federal Housing
Administration as well as the Colorado State Highway Department.   It should

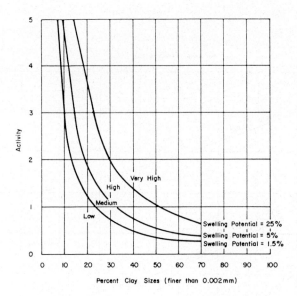

Figure 1-14. Classification chart for swelling potential. (After Seed, Woodward & Lundgren).

be pointed out that the PVC meter test in itself does not measure the swell potential.   The true swell potential of clay measured can be much greater than the indicated value.   The PVC meter test should be used only as a comparison between various swelling soils.

Ladd and Lambe [1-47] proposed a classification system in 1961 whereby soils are classified with respect to potential volume change due to both swelling and shrinkage. The method has not received wide attention.

Expansion Index.   Recently, the ASTM Committee on Soil and Rock suggested the use of an expansion index as a unified method to measure the characteristics of swelling soils.   It is claimed that the expansion index is a basic index property of soil and, therefore, is comparable to other indices such as the liquid limit, the plastic limit and the plasticity index of soil.

The sample is sieved through a No. 4 sieve. Water is added so that the degree of saturation is between 49% and 51%.   The sample is then compacted into a 40-inch diameter mold in two layers to give a total compacted depth of approximately 2 inches.   Compact each layer by 15 blows of a 5.5-lb. hammer dropping from a height of 12 inches.   Allow the prepared specimen to

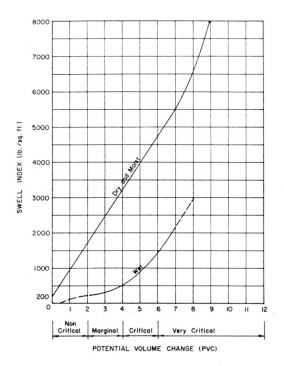

Figure 1-15. Swell index versus potential volume change. (From
"FHA Soil PVC Meter Publication," Federal Housing
Administration Publication No. 701).

consolidate under 1 lb./sq. in. pressure for a period of 10 minutes, then
inundate with water until the rate of expansion has ceased.

The expansion index is expressed as

$$EI = \frac{\Delta H}{T_1} \times 1000$$

where ΔH = Change in thickness (height)

T_1 = Initial thickness, $(T_2 - T_1)$, cm

T_2 = Final thickness, cm.

The classification of a potentially expansive soil is based on the
following table:

Expansion Index, EI	Potential Expansion
0 - 20	Very Low
21 - 50	Low
51 - 90	Medium
91 - 130	High
>130	Very High

This method offers a simple testing procedure for comparing expansive soil characteristics. However, it is limited by the use of a remolded sample. The characteristic of undisturbed soil can be very much different from that of the remolded.

Swell Index. Vijayvergiya and Ghazzaly [1-48] suggested a simple way of identifying the swell potential of clays, based on the concept of the swell index. They defined the swell index as follows:

$$I_s = \frac{W}{LL}$$

Where W = Natural water content (%)
               LL = Liquid limit (%)

The relationship between I_s and the swell potential for a wide range of liquid limit is shown on Figure 1-16.

Swell index is widely used for the design of post-tensioned slabs on expansive soil.

Direct measurement

The most satisfactory and convenient method of determining the swelling potential and the swelling pressure of an expansive clay is by direct measurement. Direct measurement of expansive soils can be achieved by the use of the conventional one-dimensional consolidometer. The consolidometer can be a platform type, scale type, or other arrangement. The load can be applied with air as in the case of the Conbel consolidometer or by direct weight as in the case of the cantilever consolidometer. The soil sample is enclosed between two porous plates and confined in a metal ring. The

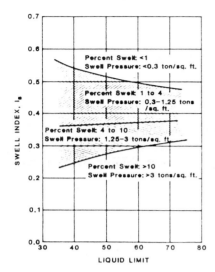

Figure 1-16.   Relationship between water content and liquid limit for
               expansive clays.

diameter of the ring ranges from 2 to 4 inches depending upon the type of
sampling device.   The thickness of the sample ranges from one-half to
1 inch.   The soil sample can be flooded both from the bottom and from the
top.   The vertical expansion measurement is reported as the percentage of
the initial height of the sample and is frequently referred to as the
percent of swell.

Such a device enables an easy and accurate measurement of the swelling
potential of a clay under various conditions.   After the soil has reached
its maximum volume increase, the sample can be reloaded and the swelling
pressure determined (chapter 2).   Thus, the swelling pressure can be
evaluated easily without resorting to devices to hold the soil volume
constant.

A great deal of data has been accumulated in files of soil engineers
and academic or governmental organizations on expanive tests using a
consolidometer.   Unfortunately, dissimilar test procedures have been used.
Thus, it is difficult to evaluate and compare the test data [1-48].   A
standardization of test procedure of a one-dimensional swell test does not
appear difficult and will salvage much of the valuable data accumulated in
the hands of the private consultants.   In the performance of a typical swell
test, the more important variables involved are as follows:

1.   State of sample.   For an undisturbed sample, this includes the
     condition of the sample, the sampling method, and the stress history of
     the sample.   For remolded samples, this includes the method of
     compaction, curing time before and after compaction, and the compaction
     density.
2.   Moisture content.   The lower the initial moisture content, the higher
     the swell.   The initial moisture content is affected by:
     (a)   The time allowed for the sample to remain in the ring before
           wetting,
     (b)   The extent of evaporation allowed while the sample is in the ring,
           and
     (c)   The temperature and humidity of the laboratory.
3.   Surcharge load.   Increasing the applied load will reduce the magnitude
     of swell.   The surcharge load for most laboratory practice ranges from
     1 to 10 psi.   Sometimes, attempts were made to duplicate the surcharge
     load with the actual footing dead load.
4.   Time allowed.   The time required to fully complete the swell process
     may vary considerably and depends on the permeability of the clay, the
     molding water content, the dry density, and the thickness of the
     sample.   For an undisturbed sample having a thickness of 1 inch, it may
     require as much as several days to complete the total available swell.

     Undoubtedly, the direct measurement method is the most important and
reliable test on expansive soils.   By standardizing the above variables, a
reliable and reproducible test can be obtained.   Also, if the concept of
swelling pressure as discussed in chapter 2 is fully understood, many of the
variables mentioned above can be simplified.

PHYSICAL PROPERTIES OF EXPANSIVE SOILS

     It is well known to soil engineers that montmorillonite clays swell
when the moisture content is increased, while swelling is absent or limited
in illite and kaolinite.   The types of soils, and the conditions under which
the most critical situation exists, can be outlined as follows:

Moisture content

     Irrespective of high swelling potential, if the moisture content of the
clay remains unchanged, there will be no volume change; structures founded
on clays with constant moisture content will not be subject to movement

caused by heaving. When the moisture content of the clay is changed, volume expansion, both in the vertical and horizontal direction, will take place. Complete saturation is not necessary to accomplish swelling. Slight changes of moisture content, in the magnitude of only 1 to 2 percent, are sufficient to cause detrimental swelling. In the laboratory, clay samples swell in the consolidometer with a slight increase of humidity. It is known that floor slabs founded on expansive soils cracked most severely when the moisture content increased slightly due to local wetting. If the floor slab is flooded, as in the case of a rising water table, the floor will heave but the extent of cracking will not be severe.

The initial moisture content of the expansive soils controls the amount of swelling. This is true both for soils in undisturbed and in remolded states. As previously discussed, the relationship between the initial moisture content and the capability of swelling has been studied by Holtz [1-42], Seed [1-43] and many others.

Very dry clays with natural moisture content below 15 percent usually indicate danger. Such clays will easily absorb moisture to as high as 35 percent with resultant damaging expansion to structures. Conversely, clays with moisture contents above 30 percent indicate that most of the expansion has already taken place and further expansion will be small. However, moist clays may desiccate due to the lowering of the water table or other changes in physical conditions and upon subsequent wetting will again exhibit swelling potential.

Dry density

Directly related to initial moisture content, the dry density of the clay is another index of expansion. Soils with dry densities in excess of 110 pcf generally exhibit high swelling potential. Remarks made by excavators complaining that the soils are as hard as a rock are indication that soils inevitably will present expansion problems.

The dry density of the clays is also reflected by the standard penetration resistance test results. Clays with penetration resistance in excess of 15 usually possess some swelling potential. In the highly expansive clay areas of Denver, penetration resistances as high as 30 are not uncommon.

Index properties

The author has accumulated years of test data on expansive soils in the Rocky Mountain area and found that it is more convenient to correlate the expansive properties with the percentage of silt and clay (-200), the liquid limit, and the field penetration resistance. Since most lightly loaded structures will exert a maximum dead load pressure of about 1,000 psf on the footings, it is realistic to use a vertical load of 1,000 psf to gauge the swelling potential. Table 1-4 is a guide for estimating the probable volume changes of expansive soils.

The simplified classification of the expansive properties can be conveniently used by engineers as a guide for the choice of type of foundation on expansive soils. For example, for soils with a low degree of expansion, spread footing type foundations can usually be used if sufficient reinforcement is provided in the foundation walls to compensate for slight movements. For soils of medium degree of expansion, individual footings or pads can be used where the dead load of the structure can be concentrated to

Table 1-4. Data for making estimates of probable volume changes for expansive soils

Laboratory and field data			Probable expansion percent total volume change	Swelling pressure, ksf	Degree of expansion
Percentage passing No. 200 sieve	Liquid limit, percent	Standard penetration resistance, blows/ft			
>35	>60	>30	>10	>20	Very high
60-95	40-60	20-30	3-10	5-20	High
30-60	30-40	10-20	1-5	3-5	Medium
<30	<30	<30	<1	1	Low

an intensity of 3,000 to 5,000 psf. For soils of high-to-very-high degree of expansion, special consideration should be given as to the foundation type. Piers with sufficient dead load pressure and enough anchorage as described in chapter 4 should be used.

Fatigue of swelling

A clay sample is subjected to full swelling in the consolidometer, allowed to desiccate to its initial moisture content, then is saturated again. This is repeated for a number of cycles. It was observed that the

soil showed signs of fatigue after each cycle of drying and wetting
[1-50]. This phenomenon has not been under full investigation. It has been
noted that pavements founded on expansive clays which have undergone
seasonal movement due to wetting and drying have a tendency to reach a point
of stabilization after a number of years. The fatigue of swelling probably
can furnish the answer. Figure 1-17 shows a typical laboratory fatigue
curve of swelling.

Fatigue of swelling was also observed by Chu [1-51] in his research on
a controlled suction test. Chu believed that if drying and wetting cycles
are repeated, the swelling during the first cycle would be appreciably
higher than that in subsequent cycles.

Popescu [1-52] studied the behavior of expansive soil with crumb
structures under wetting-drying cycles. He found that a relative
equilibrium was reached after the fifth cycle. The study indicates that
there is a drastic reduction of swelling tendency as the drying-wetting
cycle continues. However, shrinkage behavior appears to be unchanged with
repeated drying and wetting. By wetting and drying, the dry density has a

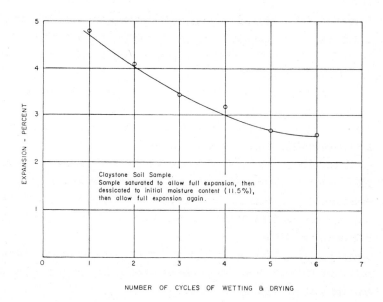

Figure 1-17. Fatigue of swelling. (After Chen, 1965).

tendency to reach the critical value, that is, the value where swelling and shrinkage equalizes. The fatigue of swelling is probably due to the gradual decrease of the dry density. When the dry density reached the critical dry density, a state of equilibrium was reached. At this point, swelling, shrinkage and dry density became stable and no further change took place.

Fatigue was also observed in practice. When seasonal wetting and drying takes place, pavement and building foundations move up and down and the movement has a tendency to reach a stable state after several wetting and drying seasons. This was observed by X.Q. Chen and Z.W. Lu [1-53].

REFERENCES

[1-1]   Holtz, W.G. and Gibbs, H.J., "Engineering Properties of Expansive Clays," Proceedings, ASCE, Vol. 80, 1954.

[1-2]   Mitchell, James K., "Fundamentals of Soil Behavior," University of California, Berkeley, 1976.

[1-3]   Donaldson, G.W., "The Occurrence of Problems of Heave and the Factors Affecting its Nature," Second International Research and Engineering Conference on Expansive Clay Soils, Texas A & M Press, 1969.

[1-4]   Krumbein, W.C., and Sloss, L.L., "Stratigraphy and Sedimentation," Second Ed., Department of Geology, Northwestern University, 1963.

[1-5]   Twenhofel, W.H., "Principles of Sedimentation," Second Ed., McGraw-Hill Book Co., Inc., 1950.

[1-6]   Aitchison, Gordon, "Status of the Art of Dealing with World Problems on Expansive Clay Soils," Engineering Effects of Moisture Changes in Soils, International Research and Engineering Conference on Expansive Soils, Texas A & M Press, 1965.

[1-7]   Hamilton, J.J., "Behavior of Expansive Soils in Western Canada," 4th International Conference on Expansive Soils, Volume II, 1980.

[1-8]   Hamilton, J.J., "Shallow Foundations on Swelling Clays in Western Canada," Engineering Effects of Moisture Change in Soils, Texas A & M University, 1965.

[1-9]   Chen, Fu Hua, "The Distribution and Characteristics of Expansive Soils in China," 5th International Conference on Expansive Soils, Adelaide, South Australia, 1984.

[1-10] "Investigation of Problems on Foundation Survey in Expansive Soils," Chinese Academy of Building Research.

[1-11] "Swelling Soils and Damage to Structures," Foundation Handbook, Chinese Construction Industry Publication Society, 1978.

[1-12] Wong, S.L., "Properties of Expansive Soils and Foundation Design," Chinese Academy of Building Research, 1980.

[1-13] Driscoll, Katkhuda and Longworth, "Geotechnical Properties of Expansive Clay Soils in Northern Jordan," 5th International Conference on Expansive Soils, Adelaide, South Australia, 1984.

[1-14] Ruwaih, I.A., "Case Studies on Swelling Soils in Saudi Arabia," 5th International Conference on Expansive Soils, Adelaide, South Australia, 1984.

[1-15] Mohan, Dinesh, Jain, G.S. and Devendra Sharma, "Foundation Practice in Expansive Soils in India," 3rd International Conference on Expansive Soils, Haifa, Israel, 1973.

[1-16] Krishan, H.R., "Building Foundation in Black Cotton Soils," Journal of the Indian National Society, New Delhi, India.

[1-17] Singh, A., "Soil Engineering in Theory and Practice," Asia Publishing House N.Y., 1967.

[1-18] Brink, A.B.A., "Engineering Geology of South Africa," Vol. 1, Building Publications Pretoria, 1979.

[1-19] Osman, M.A., and Charlie, W.A., "Engineering Properties of Expansive Soils in Sudan," 5th International Conference on Expansive Soils, Adelaide, South Australia, 1984.

[1-20] Hamid, Mohamed, E.I. "Engineering Properties of Black Cotton Soils in Sudan," Master Thesis, Colorado State University, Fort Collins, Colorado, 1978.

[1-21] Morin, W.J., and Parry, W.T., "Geotechnical Properties of Ethiopian Volcanic Soils," Geotechnique, Vol. 21, No. 3, 1971.

[1-22] Van der Merwe, C.P. and Ahronovitz, M., "The Behavior of Flexible Pavements on Expansive Soils," 3rd International Conference on Expansive Soil, Haifa, Israel, 1973.

[1-23] Driscoll, R., "A Review of British Experience of Expansive Clay Problems," 5th International Conference on Expansive Soils, Adelaide, South Australia, 1984.

[1-24] Carillo, Arnaldo, "Contribution to the Study of Expansive Clays of Peru," 2nd International Conference on Expansive Soils," College Station, Texas, 1969.

[1-25] Marquez, Edward, "Status of the Art of Dealing with World Problems on Expansive Clay Soils," Engineering Effects of Moisture Changes in

the Soils," International Research and Engineering Conference on Expansive Soils, 1965.

[1-26] Olson, Roy E., "Foundation Failures and Distress," Draft Copy for National Bureau of Standards, 1973.

[1-27] Section Study, "Properties of Expansive Clay Soils," U.S. Army Engineer Waterways Experiment Station, Vicksburg, Mississippi, 1973.

[1-28] Lum, Walter B., "Engineering Problems in Tropical and Residual Soils in Hawaii," Geotechnical Journal ASCE.

[1-29] Costa, John E. and Bilodeau, Sally W., "Geology of Denver, Colorado, United States," Bulletin of the Association of Engineering Geologists, Vol. XIX, No. 3, 1982.

[1-30] Jones, D.E. Jr., and Holtz, W.G., "Expansive Soils - The Hidden Disaster," Civil Engineering, Aug. 1973, Vol. 43, Nov. 8.

[1-31] J.H. Wiggins Co. 1978, "Building Losses from Natural Hazards: Yesterday, Today and Tomorrow." Report for National Science Foundation, Grant No. ENV.-77-08435

[1-32] Peck, R., Hanson, W. and Thornburn, T., "Foundation Engineering," John Wiley & Sons, 1974.

[1-33] Mielenz, R.C. and King, M.E., "Physical-Chemical Properties and Engineering Performance of Clays," Clay and Clay Technology, Bulletin 169, 1955, State of California, Dept. of Natural Resources.

[1-34] Tourtelot, H.A., "Geologic Origin and Distribution of Swelling Clays," Proceedings of Workshop on Expansive Clay and Shale in Highway Design and Construction, Vol. 1, 1973.

[1-35] Grim, R.E., "Clay Mineralogy," McGraw-Hill Book Co., 1968.

[1-36] Low, P.F., "Fundamental Mechanisms Involved in Expansion of Clays as Particularly Related to Clay Mineralogy," Proceedings of Workshop on Expansive Clays and Shales in Highway Design and Construction, Vol. 1, 1973.

[1-37] Warkentine, B.P., Bolt, G.H. and Miller, R.D., "Swelling Pressure of Montmorillonite," Soil Science Society of America, Proceeding 21, 1957.

[1-38] Brindley, G.W., "Identification of Clay Minerals by X-Ray Diffraction Analysis," Clays and Clay Technology, Div. of Mines Bulletin 169, 1955.

[1-39] Dodd, C.G., "Dye Adsorption as Method of Identification of Clays," Clays and Clay Technology, Div. of Mines, Bulletin 169, 1955.

[1-40] Kelley, W.P., "Interpretation of Chemical Analysis of Clays," Clays and Clay Technology, Div. of Mines, Bulletin 169, 1955.

[1-41] Ravina, I., "Swelling of Clays, Mineralogical Composition and Microstructure," Proceedings of the Third International Conference on Expansive Soils, Haifa, Israel, 1973.

[1-42] Holtz, W.G. and Gibbs, H.J., "Engineering Properties of Expansive Clays," ASCE Transactions Paper No. 2814, Vol. 121, 1956.

[1-43] Seed, H.B., Woodward, R.J. and Lundgren, R., "Prediction of Swelling Potential for Compacted Clays," Journal ASCE, Soil Mechanics and Foundations Div., Vol. 88, 1962.

[1-44] Altmeyer, W.T., "Discussion of Engineering Properties of Expansive Clays," Proceedings ASCE, Vol. 81, Separate No. 658, March, 1955.

[1-45] Holtz, W.G., "Expansive Clay - Properties and Problems," Colorado School of Mines Quarterly, Vol. 54, No. 4, 1959.

[1-46] "The character and Identification of Expansive Soils," A Report Completed for the Technical Studies Program of the Federal Housing Administration, May, 1960.

[1-47] Ladd, C.C. and Lambe, T.W., "The Identification and Behavior of Expansive Clays," Proceedings, 5th International Conference on Soil Mechanics and Foundation Engineering, Paris, Vol. I, 1961.

[1-48] Vijayvergiya, V.N., and Ghazzaly, O.I., "Prediction of Swelling Potential for Natural Clays," Proceedings of the 3rd International Conference on Expansive Clay Soils," Vol. 1, 1973.

[1-49] Kraynski, L.M., "The Need for Uniformity in Testing of Expansive Soils," Proceedings of Workshop on Expansive Clays and Shales in Highway Design and Construction, Vol. I, 1973.

[1-50] Chen, F.H., "The Use of Piers to Prevent the Uplifting of Lightly Loaded Structures Founded on Expansive Soils," Engineering Effects of Moisture Changes in Soils, Concluding Proceedings International Research and Engineering Conference on Expansive Clay Soils, Texas A & M Press, 1965.

[1-51] Chu, T. and Mou, C.H. "Volume Change Characteristics of Expansive Soils Determined by Controlled Suction Tests," Proceedings of the Third International Conference on Expansive Soils, Haifa, Israel, 1973.

[1-52] Popescu, M., "Behavior of Expansive Soils with a Crumb Structure," 4th International Conference on Expansive Soils, 1980.

[1-53] Chen, X.Q., and Lu, Z.W., "Moisture Movement and Deformation of Expansive Soils," 11th International Conference on Soil Mechanics and Foundation Engineering, Vol. 4, 1985.

Chapter 2

MECHANICS OF SWELLING

INTRODUCTION

In Chapter 1, the origin, mineralogical composition, and the basic structure of expansive soil were outlined. Obviously, if the environment of the expansive soil has not been changed, swelling does not take place. Environmental change can consist of pressure release due to excavation, desiccation caused by temperature increase, and volume increase because of the introduction of moisture. By far the most important element and of most concern to the practicing engineer is the effect of water on expansive soils.

Kraynski [2-1] stated in his review paper on expansive soils, "There must be a potential gradient which can cause water migration and a continuous passage through which water transfer can take place." With the introduction of water, volumetric expansion takes place. If pressure is applied to prevent expansion, the pressure required to maintain the initial volume is the swelling pressure.

This chapter presents a discussion on the migration of water, swelling potential, and swelling pressure, as well as lateral pressure and shrinkage.

MOISTURE MIGRATION

The pattern of moisture migration depends on the geological formations, climatic conditions, topographic features, soil types, and ground-water level. Important differences in the moisture migration pattern between covered and natural areas have been studied extensively by the Commonwealth Scientific and Industrial Research Organization in Australia. Much research has been conducted in recent years by highway organizations in Australia, South Africa, and the United States in an attempt to stabilize pavements constructed in expansive soil areas.

Moisture transfer

The most common method of moisture transfer is by gravity. The seepage of surface water, precipitation, and snow melting into the soil are common

examples.   The moisture migration can occur in all directions.   Under
artesian conditions, the flow can be upward.   In stiff clays and in shale
bedrock, the flow generally occurs in the bedding planes or follows
continuous fractures and fissures.   Shrinkage cracks which develop due to
surface desiccation provide easy access of water into the deep soils.

In fine-grained soils, capillary force is a significant means of water
transfer.   The height of water rise into the capillary fringe varies
inversely with the radius of the capillary tube.   In clean, coarse gravel,
the capillary rise is insignificant.   In clean sands, the rise is a few
inches; in fine sands, the rise is one or two feet; in silt, up to 10 to 12
feet; and in clay, a rise of more than 1000 feet is theoretically possible.

It is well recognized by observant soil engineers that the heaving of
expansive soils may take place without the presence of free water.   Vapor
transfer plays an important role in providing the means for the volume
increase of expansive soils.   Water vapor at a temperature higher than its
surroundings migrates toward the cooler area to equalize the thermal energy
of the two areas.   When water reaches the cooler area, generally the covered
area beneath a structure, condensation can take place and provide sufficient
moisture to initiate swelling.

Thermal gradients can also cause moisture migration through the liquid
phase of the soils.   Experiments conducted at Princeton University show that
a temperature differential of 1°C was at least equivalent to a hydrostatic
head of three feet in its ability to cause moisture migration.   The thermal
gradient reaches maximum efficiency when the moisture content in the soil is
near the plastic limit.   Covering the ground surface around the building
with plastic membranes creates a thermal gradient which may encourage
moisture from lawn watering to transfer to the foundation soils.

Vapor and liquid moisture transfer under thermal gradient can be an
important cause of the swelling of moisture-deficient soils.

Depth of moisture fluctuation

Kraynski [2-1] explained the moisture content variation with depth in a
homogeneous soil by Figure 2-1.   In a covered area, the moisture profile is
shown by curve 1.   There is no gain or loss of moisture to the atmosphere.
The moisture content of the soil decreases with depth.   Curve 2 indicates
the moisture content variation with depth in the same area in uncovered
natural conditions.   Evaporation causes loss of moisture in the soil near
the ground surface.   However, the influence of evaporation decreases with

depth and at some depth, H_d, the moisture content equilibrium remains the same as the covered condition. Kraynski referred to this depth as the Depth of Desiccation. The value of H_d depends on the climatic conditions, the type of soil, and the location of the water table. This depth represents the total thickness of material which has a potential to expand because of water deficiency. It is impossible to determine the value of H_d. The hotter and drier the climate, the greater the depth of desiccation. The maximum depth of H_d is equal to the depth to the water table, and the minimum depth is equal to the depth of the seasonal moisture content fluctuation described below.

During wet months, with heavier precipitation and higher humidity, the moisture content of near-surface soil increases and the moisture profile represented by curve 2 alters its shape to curve 3. The upper portion of curve 3 can extend beyond curve 1 in very wet seasons and behind curve 1 in dry seasons. The depth of seasonal moisture content fluctuation, H_s, indicated in Figure 2-1, depends on the variation of surface moisture, permeability of the soils, and climatic conditions. In areas where precipitation and evaporation are fairly constant, the H_s depth may be only a few feet. When a long drought is followed by an intense rainfall, the H_s depth can reach ten feet or more.

It should be noticed that in the above evaluation of the depth H_s, no consideration has been given to the man-made environment. The watering of lawns, planting of trees and shrubs, discharge of roof drains, formation of drainage channels and swales, and the possibility of utility line leakage all increase the value of H_s. It is not uncommon that the H_s depth can reach as much as 25 feet (see chapter 7).

When areas are covered by structures such as buildings, pavements, sidewalks or aprons, evaporation is blocked or partially retarded. The moisture content beneath the covered area increases due to gravitational migration, capillary action, vapor and liquid thermal transfer, and, in the course of several years, the depth of seasonal moisture content fluctuation H_s can approach to the depth of desiccation H_d.

The shifting of the moisture profile of a swelling soil from the natural conditions represented by curves 1 and 2 to curve 3 in Figure 2-1 for covered conditions is the cause of significant damage. Since moisture transfer is a slow process, it is not surprising that the distress of a building often takes place several years after occupancy. In the course of investigating a cracked building, it is not unusual to find that the

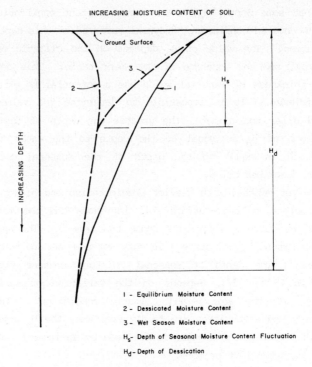

Figure 2-1.   Moisture content variation with depth below ground
surface.   (After Kraynski, 1967)

moisture content of the soils beneath the covered area or in the vicinity of
the covered area had substantially increased.

SWELLING POTENTIAL

Although the swelling phenomenon has been fully recognized for many
years, a definite method of measuring the swelling potential of clay has not
been established.

ASTM (1986) defined percent heave or settlement as percent increase or
decrease in the ratio of the change in vertical height of a column of
in situ soil.   As to the swell potential, no clear definition has been

given. Obviously, swell potential is affected by the state of the sample,
the surcharge load and the initial moisture content.

In 1962, Seed [2-2] defined swelling potential as the percentage swell
of a laterally confined sample on soaking under the 1-psi surcharge after
being compacted to maximum density at the optimum water content in the
standard AASHO compaction test. The following simplified relationship was
established:

$$S = 60K(PI)^{2.44}$$

in which S = Swell potential

and $K = 3.6 \times 10^{-5}$ and is a constant

The above equation applies only to soils with a clay content between 8
and 65 percent; the computed value is probably accurate to within about 33
percent of the laboratory-determined swell potential.

Lambe [2-3] used the swell index to measure the expansion
characteristics of clay. The swell index is defined as the slope of the
e-log p curve. A pressure increment from 1.0 to 0.1 kg/cm^2 was used.

It is clear that both definitions have their limitations. Seed's
definition is confined to remolded soil while the slope of the e-log p curve
depends on the initial moisture content as well as the surcharge pressure.
The difficulty in providing a suitable yardstick for measuring swelling
characteristics is caused by the presence of the numerous variables
involved. Consequently, to this date, when high swelling soil in San
Antonio, Texas is discussed, it is not the same standard as the swelling of
the clay shale in the Rocky Mountain area.

In another published method, the swell potential is defined as the
swell (percentage on a deformation basis) of an undisturbed specimen from an
air-dried to a saturated condition under 1 psi surcharge.

D.R. Snethen [2-4] defined swell potential as the equilibrium vertical
volume change or deformation from an oedometer-type test (i.e., total
lateral confinement) expressed as a percentage of original height of an
undisturbed specimen from its natural moisture content and density to a
state of saturation under an applied load equivalent to the in situ
overburden pressure.

From a practicing engineer's viewpoint, the above definition reflects a
meaningful concept of swell potential. As indicated by D.R. Snethen, the

definition may be amended to reflect the final stress conditions such as the applied load from the structure or the fill placement condition.

ASTM in Designation D-4546-85 defined "swell", "free swell" and "percent heave" which are also methods of testing. The standard test methods for one-dimensional swell or settlement potential of cohesive soils are given in Appendix A.

Factors affecting volume change

In the preceding chapter under the "Physical Properties of Expansive Soil," the importance of initial moisture content and initial dry density was discussed. These and other environmental conditions are extremely important in determining the amount of swell. The following factors influence the results obtained in loaded swell tests on soils of any mineralogical composition:

1. Initial moisture content. In testing undisturbed samples, care should be taken in selecting the sample with the most critical moisture content. Usually, tests should be performed on the driest sample. Frequency of testing is important so as to cover all possible conditions. For remolded samples, it is obvious that the initial moisture content controls the volume change. Field condition and construction specifications dictate the moisture requirement. In addition, a great deal of attention should be directed to the time element--the time elapsed between sampling and testing and the time elapsed when the sample is placed in the consolidometer and when the wetting takes place or when the load is applied.

2. Initial dry density. The single most important factor affecting swelling characteristics of swelling soils is density. Detailed discussions are given later under "Swelling Pressure." In remolded tests, the initial compaction condition is critical. Swell tests may decide the degree of compaction required in the placement of fill. Since moisture content and dry density are closely related, they should be examined concurrently.

3. Surcharge pressure. A small surcharge load in the range of 0.35 to 1 psi has been suggested for a seating load in the swell test. Since swell is very sensitive to changes in pressure in the lower ranges of pressure (less than about 1 psi), the use of low surcharge pressure may lead to erratic and erroneous test results. Since most footing

foundations can exert a pressure of about 1,000 psf on the soil, it is recommended that this value be used for a surcharge load.

4. Time allowed for swell. The time required for the soil to reach its maximum swell potential may vary considerably depending essentially on the initial density, permeability, and the thickness of the sample. For remolded samples, generally 24 hours is sufficient to obtain 95 percent of the total available swell. At the same time, for undisturbed high density clay shale, it may require several days or even a week before complete saturation can be achieved. For remolded samples, the initial added water must be evenly distributed. This requires a minimum curing time of six hours for reproducible results.

5. Size and thickness. Sample thickness affects the time required for total saturation. To expedite testing time, a sample thickness of less than one inch should be used. Greater thickness may introduce excessive side friction. At the same time, in a cutting from an undisturbed sample, a small thickness may introduce surface disturbance and exclude the possible effect of granular particles, fissures, and seams in the soil.

The diameter of the sample is also significant, although only vertical rise is measured in the consolidometer. The smaller the diameter, the larger the effect of side friction. Sample diameter is controlled by a sampling device. In general, the diameter should not be less than two inches. Standardization on both size and thickness of sample is necessary for comparable results.

From the preceding analysis, it is obvious that for a certain clay with known properties, the swelling characteristics vary greatly with the variation of one or more of the above environmental or placement conditions. It would be erroneous to compare the swelling characteristics of different soils without first clearly defining the placement condition.

A reliable and reproducible test which is to be considered as a basis for the classification of potential expansive soil must be standardized at least for the following environmental conditions:

For undisturbed sample:

Surcharge pressure

Size and thickness

Time required for the test

Initial moisture content and density must be qualified

For remolded sample:

Initial moisture content

Initial dry density
Method of compaction
Surcharge pressure
Size and thickness
Time required for the test
Curing time allowed

Clearly, one set of standards can hardly cover the very complicated variables involved. With the introduction of the concept of swelling pressure, the above environmental condition can be simplified and test results can be compared as later explained under "Swelling Pressure."

Total heave

The amount of total heave and the rate of heave on which a structure is founded are very complex. Unlike settlement prediction, the heave estimate depends on many factors which cannot be readily determined. The following basic factors should be considered:

1. Climate. Climatic conditions involving precipitation, evaporation and transpiration affect the moisture in the soil as previously discussed under "Moisture Transfer." The depth and degree of desiccation affect the amount of swell in a given soil horizon. Climatic conditions partially affect the desiccation.

2. The thickness of expansive soil stratum. In most cases, the thickness of the expansive soil stratum extends down to a great depth, and the practical thickness is governed by the surface water penetration into the stratum. For practical purposes, we assume a depth of 15 feet, and in some extreme cases, surface water penetrates through the seams and fissures of expansive clays to as much as 30 feet. In this case, the thickness of the expansive soil stratum is controlled by the depth to water table.

3. The depth to water table. Soil below the water table should be in a state of complete saturation; consequently, no swelling of clay should take place for the portion of soil below the water table. However, in the case of claystone bedrock, due to its fine-grained structure and its nearly impermeable nature, water is not able to penetrate and saturate the material. The only access of water into the stratum is through its seams and fissures. Consequently, when taking the moisture content of submerged claystone, one may find very erratic values. Near saturation conditions can be found in the vicinity of seams and

fissures, while relatively dry conditions can be found elsewhere. Under a fairly stable and unchanging environment, the claystone portion of bedrock below the water table can be considered to be free from a volume change. Prediction of total heave can be considered only on the stratum thickness above the water table.

Changes of environment can, however, alter the entire picture. Construction operations such as pier drilling can break through the system of interlaced seams and fissures in the claystone structure, allowing water to saturate the otherwise dry area, thus allowing further swelling of the otherwise stable material.

Fluctuation of the water table is also a factor contributing to a change of environment. When the water table drops and subsequently rises again, the flow of water need not follow the existing paths. When water flows along new paths, swelling again may take place.

Therefore, in considering the thickness of an expansive soil stratum, consideration should be given not only to the thickness above the water table but also to the depth below the water table. This depth should be at least equal to that of the possible fluctuation of the water table.

4. The nature and degree of desiccation of the soil. The predicted amount of heave depends on the initial condition of the soil immediately after construction. If the excavation is exposed for a long period of time, desiccation takes place, and upon subsequent wetting, more swelling may take place. In the laboratory, we conducted a swell test both in its natural moisture content and in an air dried condition. Air dried soils swell much more than soils in their natural state. It is difficult to predict what will actually take place, but it is possible to give a maximum possible amount and a minimum likely amount.

5. The initial stress condition in the soil. The single most important element controlling the swelling pressure as well as the swell potential is the in situ density of the soil. On the completion of excavation, the stress condition in the soil mass undergoes changes. There is elastic rebound. Stress release increases the void-ratio and alters the density. Such physical changes, however, do not take place instantaneously. If construction proceeds without delay, the structural load compensates for the stress release. We do not believe this is a significant amount.

6. Permeability and rate of heave. The permeability of the soil determines the rate of ingress of water into the soil either by

gravitational flow or diffusion, and this in turn determines the rate of heave. The higher the rate of heave, the more quickly the soil responds to any changes in the environmental conditions, and thus the effect of any local influences is emphasized. At the same time, the higher the permeability, the greater the depth to which any localized moisture penetrates, thus engendering greater movement and greater differential movement. Therefore, the permeability is an important factor; the higher the permeability, the greater the probability of differential movement.

7.   Extraneous influence. The above mentioned basic factors, although difficult to predict, can still be evaluated theoretically. Extraneous influence at the same time is totally unpredictable. The supply of additional moisture accelerates heave, for instance, if there is an interruption of the subdrain system to allow the sudden rise of a perched water table. The development of the area, especially residential construction, contributes to a drastic rise of a perched water table.

Various methods have been proposed to predict the amount of total heave under a given structural load. These are the double oedometer method [2-6], the Department of Navy Method [2-7], the South African method [2-8] and the Del Fredlund method [2-9]. All suggested methods have limitations. The double oedometer method developed by Jennings and Knight is based on the concept of effective stress and has received wide attention. The general test procedure is given as follows:

Two consolidometer rings are filled with undisturbed samples from adjacent locations. The first sample is kept at its natural moisture content and a confined compression test is performed. The second sample is flooded with water at a low pressure of 20 psf. After the sample is fully wetted, a consolidation test is performed in the conventional manner. The two compression curves are plotted on the same diagram and one of the curves is selected for vertical adjustment so as to coincide with the virgin sections of the curves as shown on Figure 2-2.

A soil sample taken at depth z has an overburden pressure $P_o = \gamma z$ where γ is the density of the soil. The void-ratio at the overburden pressure is e_o. Settlement due to load increment ΔP can be calculated from the change of void-ratio $e_o - e_1$. If no load is applied, the soil under a covered area gains moisture and swelling takes place. The condition alters P_o, resulting in e_o having a new effective pressure $P_o + U_L$ represented in the upper saturated curve by $P_o + U_L$ and e_2. If D is the depth to the water

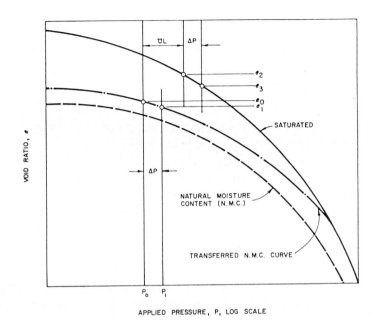

Figure 2-2. Log P curves showing adjustment to bring straight line
portions coincident. (After Jennings & Knight, 1958)

table and γ_w is the unit weight of water, $U_L = \gamma_w (D - z)$. The effect of
the load increment is again taken into consideration and the final values
are $(P_o + U_L + \Delta P$ and $e_3)$. The final conditions of movement may then be
predicted by adding the void-ratio changes over the whole profile.

The double oedometer method is based on the assumption that there is a
point during compression at which the initially unsaturated soils pass from
an applied pressure to an effective phenomenon and the compression curve
joins with the virgin consolidation curve. Typical heave calculations are
shown on Table 2-1.

The Department of Navy in its Design Manual [2-7] outlined a procedure
for estimating the magnitude of swelling that may occur when footings are
built in expansive soils. Figure 2-3 also indicates a method of determining
the necessary undercut to reduce the heave to an acceptable value. The
shortcome of this procedure is that it is assumed that at a lower depth the

Table 2-1. Results of heave calculations

1	2	3	4	5	6	VOID RATIO				11	12	13
						7	8	9	10			
Layer No.	Depths at top and bottom of layer feet	Depth of double Oedom. test feet	P_o at mean depth ton/sq.ft.	ΔP at mean depth ton/sq.ft.	U_L at mean depth ton/sq.ft.	e_0	e_1	e_2	e_3	Heave of unloaded surface $H\frac{(e_0-e_2)}{1+e_0}$ In.	Settlement under buildings. Ins. $H\frac{(e_0-e_1)}{1+e_0}$ In.	Net heave under building $H\frac{(e_0-e_3)}{1+e_0}$ In.
1	4.00-6.56	5.25	0.30	0.10	0.97	0.419	0.418	0.492	0.481	1.585	0.021	1.342
2	6.56-8.87	7.87	0.45	0.07	0.89	0.428	0.427	0.460	0.455	0.623	0.019	0.525
3	8.87-10.80	9.87	0.59	0.06	0.84	0.418	0.417	0.449	0.447	0.506	0.016	0.474
4	10.80-13.35	11.74	0.73	0.04	0.78	0.544	0.543	0.563	0.562	0.362	0.020	0.343
5	13.35-18.81	14.85	0.97	0.04	0.70	0.626	0.625	0.640	0.639	0.565	0.040	0.524
6	18.81-18.87	18.87	1.21	0.04	0.59	0.523	0.522	0.524	0.523	0.017	0.017	0
									TOTALS:	3.658	0.133	3.208

(After Jennings & Knight, 1958)

soil does not swell as much as at an upper depth due to overburden pressure. In fact, even at a depth of eight feet the overburden pressure amounts to only, at most, 1 kip/ft.^{2} which is small in highly expansive soils. For slab-on-grade construction on highway pavement where there is little weight of structure, this method cannot be applied.

Probably, the nearest practical approach to this problem is that of the Van der Merwe method [2-8].

The Van Der Merwe method, commonly known as the South African Method, started by classifying the swell potential of soil into very high to low categories as shown in Figure 2-4. Then assign potential expansive (P.E.) expressed in in./ft. of thickness based on the following:

Swell Potential	Potential Expansion (P.E.) In./ft.
Very high	1
High	1/2
Medium	1/4
Low	0

Assume the thickness of an expansive soil layer or the lowest level of ground water. Divide this thickness (Z) to several soil layers with variable swell potential.

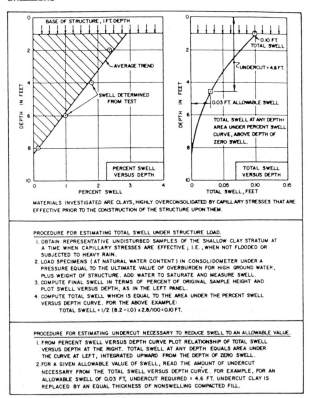

Figure 2-3. Computation of swell of desiccated clays

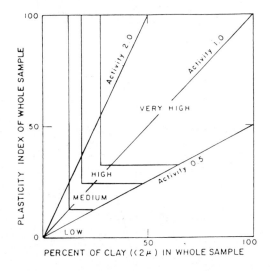

Figure 2-4. Volume change potential classification for clay soils

The total expansion can be expressed by

$$\Delta H_S = \sum_{e=1}^{n} \Delta_e$$

Where ΔH_S = total expansion (in.)

Δ_e = (P.E.)(ΔD)(F)

$F = Log^{-1} (- \dfrac{D}{2})$

Z = total thickness of expansive soil layer (ft.)

D = depth to mid point of each layer (ft)

ΔD = thickness of individual layer (ft)

Figure 2-5 is a typical example of calculation.

EXAMPLE:

LAYER	THICKNESS (FT) ΔD	P.E.	D (FT)	F	Δ_e (INCHES)
I	5	0	2.5	0.75	0
2	8	1	9	0.35	2.80
3	2	1/2	14	0.20	0.20
4	5	1	17.5	0.13	0.65
					$\Sigma = 3.65$

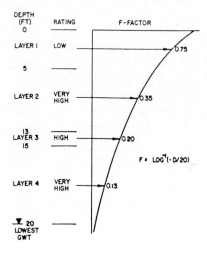

Figure 2-5. Estimating swell using the South African method

Del Fredlund [2-9] used the results of oedometer tests in terms of void-ratio to predict the total heave. The heave in an individual soil layer can be written as follows:

$$\Delta h_i = h_i \left[\frac{\Delta e}{1 + eo}\right]$$

Δh_i = heave in a layer

h_i = thickness of the layer under consideration

Δ_e = change in void ratio

     = $e_f - e_o$

e_f = final void ratio

e_o = initial void ratio

Total heave $\Delta h = \sum \Delta h_i$

$$\Delta h_i = h_i \frac{C_s}{1 + e_o} \log P_f/P_o$$

C_s = swell index, slope of e-log P curve

P_f = final stress state

P_o = initial stress state

Figure 2-6 indicates an example of the total heave calculation.

Consider a 6.5-foot-thick layer of expansive soil. The initial void ratio of the soil is 1.0, the total unit weight is 114.6 pcf and the swelling index is 0.1. Assuming that no correction on swelling pressure is made, the swelling pressure from the oedometer test is 4174 psf, and the total predicted heave is on the order of 4.5 inches.

Many empirical methods have been established to predict heave. The input data commonly needed consists of classification data such as the Atterberg limits, initial water content, dry density and percent clay content.

Johnson and Snethen [2-10] introduced the suction method for heave prediction. They concluded that the suction method is simple, economical, expedient and capable of simulating field conditions. They also summarized the empirical methods as shown in the following Table [2-2].

The search by various investigators for a reliable method for predicting the total heave is probably affected by the concept of ultimate settlement in the theory of consolidation. For many years, engineers have

been familiar with the calculation of ultimate settlement and differential
settlement of a structure founded on clay, and it is assumed that the total

Table 2-2. Empirical methods for predicting heave

References	Description
McDowell [2-11]	A procedure based on swell test results of compacted Texas soils. Field heave is estimated from a family of curves using the Atterberg limits, w_o, and q for each soil stratum. The value of w_o is compared to maximum $(0.47LL + 2)$ and minimum $(0.2LL + 9)$ w's.
Seed et al. [2-12]	$S_p = 0.00216PI^{2.44}$ for soil compacted at optimum w and maximum density to saturation and 6.90 kPa (1 psi) q.
Van Der Merwe [2-8]	$\Delta H_s = \sum\limits_{e=1}^{n} (P.E.)\ (\Delta D)\ Log^{-1}\ (-\dfrac{D}{2})$
Vijayvergiya and Ghazzaly [2-13]	$Log\ S_p = 1/12\ (0.44LL - w_o + 5.5)$ from w_o to saturation for q = 0.1 tsf (10 kPa).
Vijayvergiya and Sullivan [2-14]	$Log\ S_p = 0.0526\ \gamma d + 0.033LL - 6.8$ from w_o to saturation for 6.9 kPa (1 psi). The value of γd is given in lb/ft^3; $1\ lb/ft^3 = 0.016\ Mg/m^3$. Equation derived herein from the given family of curves.
Nayak and Christensen [2-15]	$S_p = (0.0229PI \cdot 1.45C)/w_o + 6.38$ for compacted soil and 6.9 kPa (1 psi) q to saturation.
Schneider and Poor [2-16]	$Log\ S_p = 0.9\ (PI/W_o) - 1.19$ for no fill or weight on the swelling soil to saturation.
McKeen [2-17]	A procedure relating soil suction to S_p including effect of q. Requires use of graphs, shrinkage limit, PI, LL, C, and estimates of $\gamma\ m0°$ and $\gamma\ mi°$.
Johnson [2-18]	For PI $\geq$ 40, $S_p = 23.82 + 0.7346\ PI - 0.1458H - 1.7w_o + 0.0025PIw_o - 0.00884PIH$, for PI $\leq$ 40, $S_p = -9.18 + 1.5546\ PI + 0.08424H + 0.1\ w_o - 0.0432PIw_o - 0.01215\ PIH$ for q = 6.9 kPa (1 psi) to saturation.

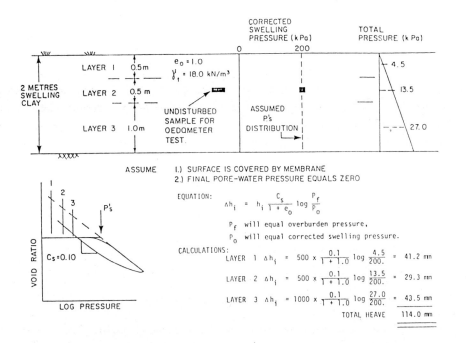

Figure 2-6. Calculations for total heave.

heave can also be predicted. There are some fundamental differences between the behavior of settling and heaving soil. Some of them are as follows:

1. Settlement of clay under load takes place without the aid of wetting, while expansion of clay is not realized without moisture increase.

2. The total amount of heave depends on the environmental condition, such as the extent of wetting, the duration of wetting, and the pattern of moisture migration. Such variables cannot be ascertained, and consequently, any total heave prediction can be entirely erroneous.

3. Differential settlement is usually described as a percent of the ultimate settlement. However, in the case of expansive soils, one corner of the building may be subjected to maximum heave due to excessive wetting while another corner may have no movement. Therefore, in the case of swelling soils, differential heaving can

equal the total heave. No correlation between differential and total
heave can be established.

Effective stress

Terzaghi developed the principle of effective stress in the early
1920's. It is believed that the principle can be applied to all soil
behaviors. In general terms, the principle of effective stress states that
the behavior of an element of soil mass depends not on the total stress
applied to the element but rather on the difference between this total
stress and the stress present in the pore fluid. For saturated soil, the
effective stress is defined as:

$$\bar{\sigma} = \sigma - \mu$$

where: $\bar{\sigma}$ = effective stress

σ = total normal stress

μ = stress in pore fluid or pore water pressure

The preceding equation is based on saturated soils. For partially
saturated soils, consideration should be given to the fact that pore
pressure acts only on part of the area of any plane through the soil. In
addition, an electro-chemical force of attraction and repulsion force may
act through the spaces not filled with water. Consequently, the effective
equation should be modified by adding the vector sum of attractive forces
and subtracting the vector sum of repulsive forces.

Aitchison and Richards [2-19] stated, "It is not only possible but
essential for the effective stress principle to be used in the
quantification of expansive soil behavior."

Lambe [2-3] stated in 1959 after making in-depth research on the
application of the concept of effective stress to explain the swelling soil
behavior, "One wonders whether, for example, we are hindering our
understanding of the nature of the extremely plastic swelling soils by
forcing them to fit the effective stress concept."

The author shares the thinking of Lambe and believes that the mechanics
of expansive soil are totally different from the theory of consolidation and
the shearing strength of soils. It should be considered as a new phase of
soil mechanics.

SWELLING PRESSURE

ASTM defines swelling pressure as the pressure which prevents the specimen from swelling or that pressure which is required to return the specimen back to its original state (void ratio, height) after swelling. Essentially, the methods of measuring swelling pressure can be either stress controlled or strain controlled.

Stress controlled

For stress controlled tests, the conventional oedometer is used. The samples are placed in the consolidation ring trimmed to a height of 3/4 to 1 inch. The samples are subjected to a vertical pressure ranging from 500 psf to 2,000 psf depending upon the expected field conditions. On the completion of consolidation, water is added to the sample. When swelling of the sample has ceased, the vertical stress is increased in increments until it has been compressed to its original height. The stress required to compress the sample to its original height is commonly termed the zero volume change swelling pressure. A typical swell-consolidation curve is shown on Figure 2-7.

In China, geotechnical engineers determine swelling pressure based on the unloading curve instead of the compression curve. [2-20]

Initial void ratio as shown on Figure 2-8 is e_o. Under pressure, it decreases to e_1. Curve AB indicates a compression curve. Under pressure p_1 the sample is wetted and expansion takes place. When volume change has stabilized, void ratio increases to e_2. Continue unloading until zero pressure is registered. The unloading should take place in increments. Each increment should take place when the expansion has been completed and stabilized. Thus, Curve BC and CD can be plotted. A horizontal line is drawn through A intersected curve CD at point N.

From Figure 2-8 it is seen that below line A-A', all points on Curves AB, BC and CN have a void ratio smaller than e_o. This indicates that as long as the applied pressure is higher than P_N, compression takes place. In other words, when a pressure higher than P_N is applied on the sample and upon wetting, the end result is compression instead of expansion. P_N is, therefore, the swelling pressure.

Strain controlled

The strain controlled method is based on the principle of controlling
the strain that is developed as water is added.

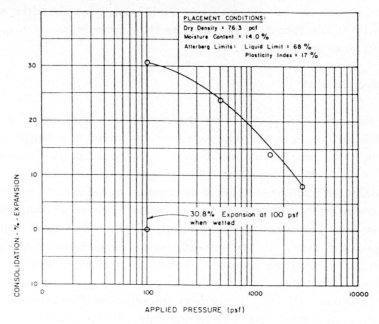

Figure 2-7. Typical stress control swell-consolidation curve.

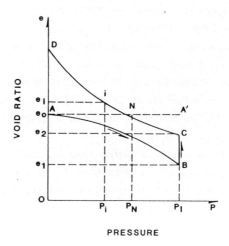

Figure 2-8. Typical swell-consolidation curve based on unloading condition.

In the strain controlled test, modification of the conventional oedometer is required to allow the control of strain during testing and measurement of the resulting loads. Figure 2-9 shows one type of controlled strain apparatus developed by Porter & Nelson. [2-22].

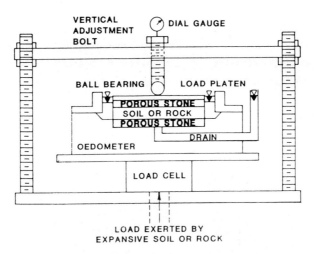

Figure 2-9. Modified consolidometer for strain controlled test.

A load frame was filled with a vertical adjustment bolt to control the strain. A stiff load cell was mounted at the bottom to measure the load resulting from swelling during imbibation of water.

A similar strain control test was made by El-Fatih Al in India [2-21]. He used proving rings to register the swelling pressure and continually adjusted them to keep the soil sample at its original volume.

Comparison

Porter & Nelson [2-22] concluded that the agreement between consolidation-swell test results and those from controlled strain tests was not exact for all samples but was very close.

El-Fatih Ali concluded that the constant volume method gives results of swelling pressure that are generally smaller than those obtained from the oedometer method.

Theoretically, the author believes that the swelling pressure is a built-in soil property similar to that of the index property. Defining swelling pressure, either by stress control by loading curve, stress control by unloading curve or by strain control method, the result should be the same.

D.G. Fredlund's research [2-9] indicated that laboratory determination of swelling pressure may not represent the actual swelling pressure in the field and should be corrected.

The corrected swelling pressure consists of:

1.   Adjustment should be made to the laboratory data in order to account for the compressibility of the oedometer apparatus.

2.   Correction must be applied for sampling disturbance.

It is possible for the "corrected" swelling pressures to be more than 300 percent of the "uncorrected" swelling pressure.

The "corrected" swelling pressure is designated as the intersection of the bisector of the angle formed by these lines and a line tangent to the curve which is parallel to the slope of the rebound curve as shown in Figure 2-10.

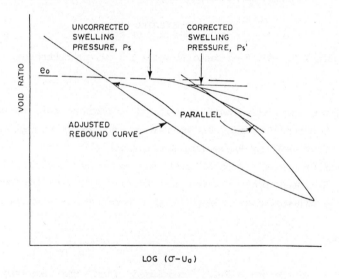

Figure 2-10.   Construction procedure to correct for the effect of sampling disturbance.   (After D.G. Fredlund).

MECHANICS OF SWELLING PRESSURE

In the course of the last 30 years, thousands of swell tests were conducted by the author on various kinds of expansive soils found in the Rocky Mountain area. The standard procedure used to conduct these tests was to place the undisturbed sample in a consolidometer under a surcharge load of 1,000 psf (about 7 psi) for 24 hours, saturate the sample, measure and record the amount of volume change.

The amount of volume change exhibited by various soils under various placement conditions varies greatly. Furthermore, it was found that soil obtained from beneath a structure that has undergone severe foundation movement may not possess high swell potential. It was suspected that there must be a single soil property that governs the swelling characteristics. After the sample swelled to its maximum extent, the specimen was loaded until it returned to its initial volume and the pressure required to do this was designated as swelling pressure.

It is suspected that the swelling pressure is the built-in property of expansive soil and is not affected by placement condition or environmental condition.

Test procedure

The clay selected for making this study is a claystone shale typical of those found in southeast Denver. Such soil has caused a great deal of damage to lightly loaded structures such as residential houses. Most of the houses in this area are founded with piers bottomed in a zone where a change of moisture content is unlikely to take place; however, slabs placed on such soil have experienced generally severe movement. Uplift movement in excess of six inches is not uncommon.

The physical properties of such clay are as follows:

Liquid limit	44.4%
Plasticity index	24.4%
Shrinkage limit	14.5%
Sand	0%
Silt	63.0%
Clay (percent smaller than 0.005 mm)	37.0%
Optimum moisture content	19.5%

Maximum dry density (standard
    Proctor test)   108.4 pcf
Free swell (USBR method)   75.0%
Specific gravity   2.67

In addition to the above, the mineral content, in percent, of such clay is
as follows:

Montmorillonite   25.0
Calcite   5.0
Quartz   25.0
Feldspar   10 to 25.0
Kaolinite   5.0

Because of the erratic formation of natural soil strata, for research
purposes, it is necessary to use only remolded samples so that the placement
condition can be duplicated.

The air dry sample is prepared by passing it through a No. 40 sieve.
Moisture content is added to the air dry clay and then allowed to age in a
sealed container for a period of 48 hours. The sample is compacted in a
two-inch-diameter, one-inch-thick consolidometer ring in three layers with
predetermined compaction effort. The required density is obtained as near
as possible by an experienced technician. It should be noted that in the
test results presented, some deviation of density took place which resulted
in some erratic test results.

Surcharge pressure

It is a well recognized fact that if sufficient load is applied on an
expansive clay, the detrimental volume increase can be controlled. The
surcharge pressure applied to the soil sample in the consolidometer
simulates the dead-load pressure exerted on the footings or pier
foundation. Figure 2-11 and Table 2-3 indicate that with a surcharge
pressure of 1,000 psf, upon wetting the clay swelled 5.9 percent with a
swelling pressure of 12,000 psf. By increasing the surcharge pressure to
5,000 psf, the amount of volume increase was limited to 1.6 percent, but the
swelling pressure remained unchanged.

Table 2-3. *Effect of varying pressure on volume change and swelling*
pressure for coantant density and moisture content samples.

Applied pressure (psf)	Moisture content, percent		Initial density, pcf	Volume increase, percent	Swelling pressure, psf
	Initial	Final			
1,000	11.90	24.58	105.58	5.90	12,000
2,000	11.90	25.08	106.08	3.90	13,000
3,000	11.90	24.94	105.96	2.80	12,000
5,000	11.90	25.02	105.90	1.60	12,500
7,000	11.90	25.49	105.96	1.00	12,500
Average	11.90	25.02	105.95		12,400

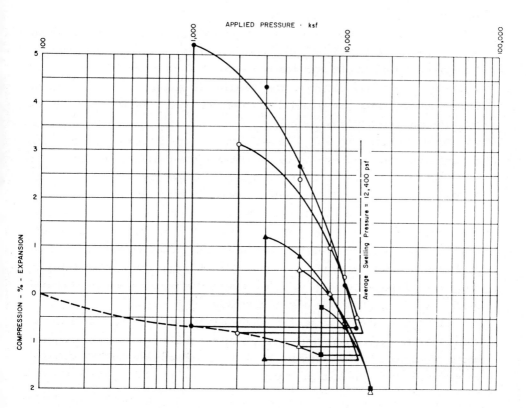

Figure 2-11. Relationship between surcharge pressure and volume increase
for constant density and moisture content samples.

Figure 2-12 indicates the relationship between volume change with surcharge pressure. These curves have a hyperbolic shape and the intersection of the curves with the abscissa indicates the pressure required for zero volume change. This pressure by definition is the swelling pressure.

The surcharge load is essential to control foundation movement. With the swelling pressure determined, a reasonable foundation design can be approached. If the swelling pressure is not excessive, on the order of 5,000 psf, a spread footing foundation can be used. The requirement is to assign a minimum dead-load pressure of 5,000 psf so that the volume change of the soil is not allowed even in excessive wetting condition. In the design of the footing foundation, it may be possible to allow certain amounts of uplift movement so as to minimize the required dead-load pressure. Uplift movement can be tolerated in certain structures in the same manner as some settlement can be tolerated in most structures. A differential uplift of three-fourths of an inch generally is considered to be tolerable. With 3/4-inch allowable differential uplift, the required dead-load pressure can be drastically reduced. For a rational design, it is advisable to actually determine the swelling pressure. From the pressure versus volume change curve and the tolerable uplift, a working swelling pressure can be established.

For highly swelling soil with swelling pressure in excess of 5,000 psf, a pier foundation is required to concentrate the dead-load pressure on a small area. It is not difficult to assign dead-load pressure in excess of 20,000 psf in a small diameter pier. However, attention must be given to the additional swelling effect on the shaft of the pier embedded in swelling soil. The swelling pressure exerted on the shaft of the pier can be many times greater than the pressure exerted at the bottom of the pier. Dead load pressure alone generally is not sufficient to prevent the uplifting of the pier. Anchorage of the pier in a zone not affected by moisture change should be used to assist the dead-load pressure requirement.

Since the volume change for surcharge pressure of one psi and the volume change under a surcharge pressure of ten psi varies considerably, the merit of using swelling pressure as a direct measurement of the swelling characteristics can be seen at once.

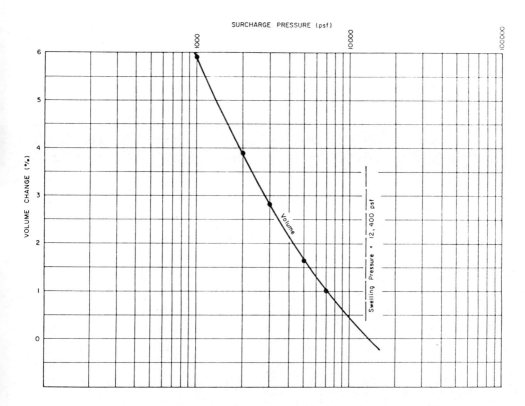

Figure 2-12. Effect of varying pressure on volume change for constant
              density and moisture content sample.

Degree of saturation

     This series of tests was performed to study the effect of the duration
of wetting on swelling characteristics. It is a well established phenomenon
that prolonged wetting results in more damage to a structure than short
duration wetting.

     Since it is difficult in a short sample height to control the duration
of wetting, in order to achieve the same effect in the laboratory the degree
of saturation on the sample was varied. The samples were compacted in the
consolidometer with uniform density and moisture content, and then a
measured amount of water was introduced into the sample. Sufficient time
was allowed for all added water to soak into the sample. The amount of

Table 2-4. *Effect of varying degree of saturation on volume change and*
swelling pressure for constant density and moisture content
samples.

Moisture content, percent		Initial density, pcf	Volume increase, percent	Swelling pressure, psf	Degree of saturation, percent
Initial	Final				
9.66	13.07	106.6	1.83	16,000	61.0
9.66	14.53	106.0	3.35	15,500	67.0
9.66	17.58	105.6	4.35	12,000	82.0
9.66	18.50	106.7	5.53	17,000	86.3
9.66	19.93	105.9	6.25	15,000	93.0
Average 9.66		106.2		15,100	

volume change and the swelling pressure recorded are shown on Table 2-4.
From Figure 2-13, it can be seen that the amount of volume change increased
in direct proportion to the degree of saturation at the end of test.

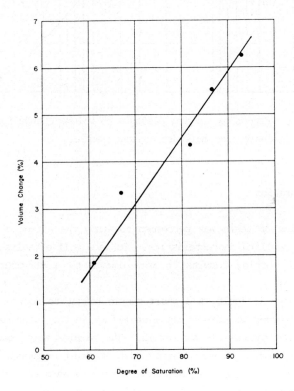

Figure 2-13. Effect of varying degree of saturation on volume change
for constant density and moisture content sample.

Figure 2-14 indicates that the swelling pressure is constant, or the pressure required to maintain constant volume is independent of the duration of wetting or the degree of saturation.

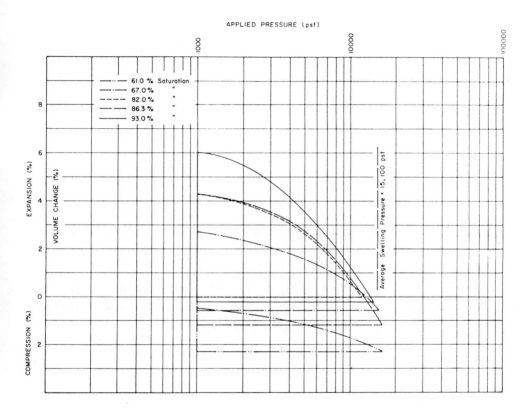

Figure 2-14. Relationship between degree of saturation and volume increase for constant density and moisture content samples.

Complete saturation is not required to result in a large volume change. There is a misconception that by removing free water the swell can be controlled. It has been a common practice to install drain tiles around a building in an attempt to remove free water and to stop foundation movement. A subdrain does not arrest the migration of moisture. Consequently, swelling can be substantial.

Without a superimposed load, the swelling of the soil cannot be controlled even with a minimum amount of moisture change. Since the swelling pressure remains constant, a short duration of wetting can cause equally heavy damage to lightly loaded structures as long duration wetting. This is the reason why it is so difficult to control slab movement for slab-on-ground construction.

Initial moisture content

Expansive soils are not subject to volume change unless there is an increase in moisture content. A drier soil swells more than a wet soil. In a series of tests, an attempt was made to determine the effect of increasing the initial moisture content on the volume change as well as swelling pressure. Figure 2-15 shows the results of a series of tests indicating the amount of volume change of soil samples compacted at constant density but varying the moisture content. As expected, the soils with low initial moisture content swelled most. Thus, the slope of the e-log p curve decreased as the initial moisture content increased. However, the swelling pressure required for zero volume change remained practically constant.

Kassiff & Baker in 1971 [2-23] stated that if clay is given enough time for aging, that for the same dry density, the swell pressure is not affected by the moisture content. The results shown in Table 2-5 confirm this

Table 2-5. *Effect of varying moisture content on volume change and swelling pressure for constant density samples.*

Initial density, pcf	Moisture content, percent		Volume increase, percent	Swelling pressure, psf
	Initial	Final		
106.97	5.84	20.34	7.71	9,500
105.93	9.95	20.77	5.55	9,500
106.27	10.77	18.75	5.03	12,500
105.60	12.48	22.09	4.30	9,500
106.47	12.92	20.54	3.48	9,000
106.37	14.84	19.59	3.30	10,500
105.46	17.97	18.50	2.15	7,000
105.73	18.59	19.41	1.38	7,500
106.35	19.37	20.18	0.75	9,000
Average 106.13		20.02		9,333

statement.   Table 2-5 indicates that the swelling pressure for the various
moisture contents ranges from 7,000 to 12,500 psf.   Due to variation in
laboratory controlled conditions,   the   initial   density   is   not   entirely
constant, and some variation in swelling pressure occurs.   For all practical
purposes, however, the swelling pressure is a constant value.   Figure 2-16
indicates the variation of moisture content versus volume change.   These
tests indicate that with moisture content slightly higher than optimum
moisture content, the volume change should be negligible.

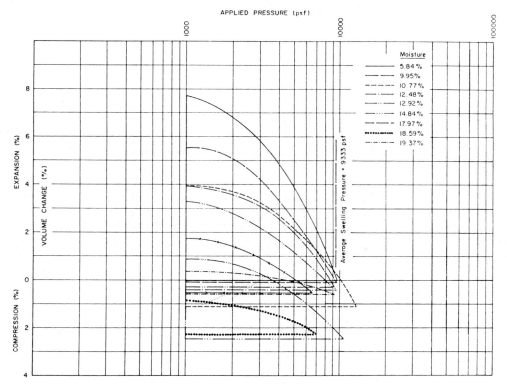

Figure 2-15.   Relationship between initial moisture content and volume
               increase for constant density samples.

The results of the above laboratory tests indicate that the increase of
the moisture content of an expansive soil is not a positive method of
controlling the expansion of the soil.   Even with high moisture content,
footings founded on swelling soil experience the same swelling pressure, and

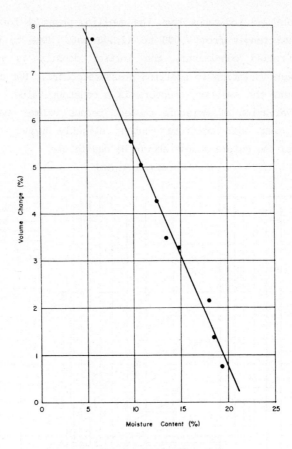

Figure 2-16.   Effect of varying moisture content on volume changes from
               constant density samples.

the same amount of dead-load pressure is required to ensure zero volume
change.   High moisture content soils experience less uplift, but the
pressure required to maintain a constant volume is not altered.   This also
indicates that the commonly accepted procedure of prewetting the foundation
excavation to eliminate the swelling characteristics is not a reliable
procedure.   Wetting of the foundation soil, if it can be accomplished, can
only serve to decrease the amount of swelling. A foundation placed on such
soil still requires the same amount of dead-load pressure.

     Soil engineers are often deceived by the low volume change of a high
moisture content soil.   If such soils possess a high swelling pressure, they

cause severe damage to structures if allowed to dry and are subsequently
wetted.

Stratum thickness

Laboratory research has been further extended to explore the effect of
stratum thickness on the amount of volume change and swelling pressure. In
this series of tests, the sample thickness ranged from 1/2 to 1 1/2
inches. Again, the samples were compacted to uniform moisture content and
density and sufficient time was allowed for complete saturation of the
thickest sample.

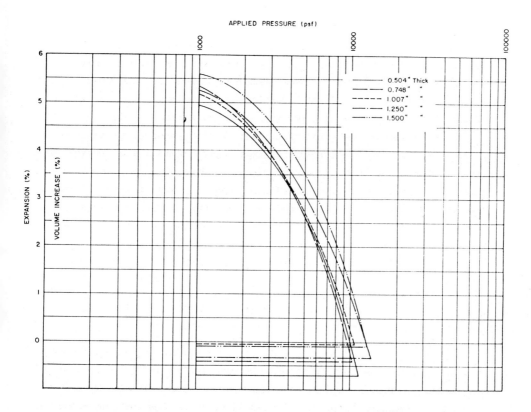

Figure 2-17.   Relationship between sample thickness and volume increase
               for constant density and moisture content samples.

Table 2-6. Effect of varying sample thickness on volume change and swelling pressure for constant density and moisture content sample.

Initial density, pcf	Moisture content, percent		Sample thickness, in.	Volume increase, percent	Volume increase. in.	Swelling pressure, psf
	Initial	Final				
105.20	10.10	22.30	0.504	5.66	0.0285	11,000
106.33	10.10	20.92	0.748	5.75	0.0430	11,500
105.31	10.10	21.14	1.007	5.15	0.0520	11,000
106.05	10.10	20.49	1.250	5.60	0.0700	15,000
106.05	10.10	20.58	1.500	5.60	0.0840	12,500
Avg. 105.78	10.10	21.08		5.54		12,200

As could be predicted from the results shown in Figure 2-18 the magnitude of the volume change is proportional to the sample thickness with the percentage of volume increase remaining constant. The shape of the e-log p curve remains almost identical for various sample thicknesses (Figure 2-17) and the swelling pressure is constant (Table 2-6).

This series of tests indicates that if the weight of a structure is capable of exerting pressure to various depths beneath the footing with equal intensity, then the volume increase can be arrested. Unfortunately, dead-load pressure exerted on the footing can only control volume change of the near surface soils. At lower depths, pressure exerted on the footing is distributed over a larger area and is not effective in preventing volume change. Deep seated swelling is controlled only by the weight of the overburden soil and not by dead-load pressure exerted on the foundation system.

Initial density

Initial density, whether undisturbed or remolded, is the only element that affects the swelling pressure. As seen in Figure 2-19 and Table 2-7 for constant moisture samples, the volume change increases with dry density, as does the swelling pressure. The family of curves has the same shape and are approximately parallel to each other. A similar relationship was found by Kassiff and Shalom [2-24]. Figure 2-20 establishes a straight line relationship between dry density and volume change. The relationship between the dry density and swelling pressure can be plotted either in

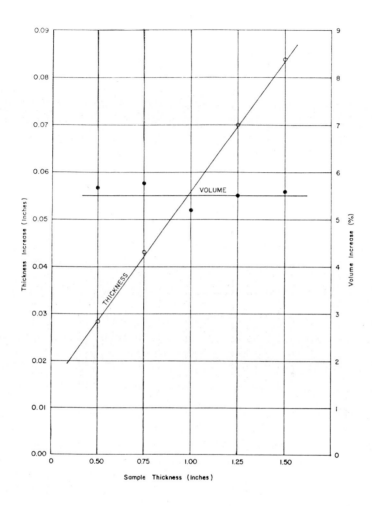

Figure 2-18.   Effect of varying thickness on volume change for constant
               density and moisture content samples.

semi-log scale or in rectangular scale.   For the semi-log scale, the curve
is a straight line as shown in Figure 2-21.

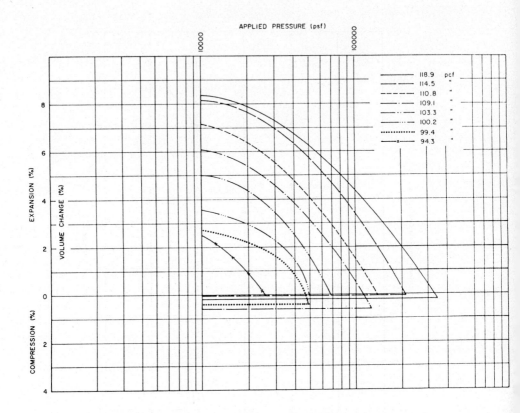

Figure 2-19.   Relationship between density and volume increase for constant
                initial moisture content samples.

The curve can be expressed as:

$$\log y = ax - b$$

where: y = swelling pressure,

           x = dry density, and

           a and b = constants depending on soil property
           and "a" is the slope of the curve.

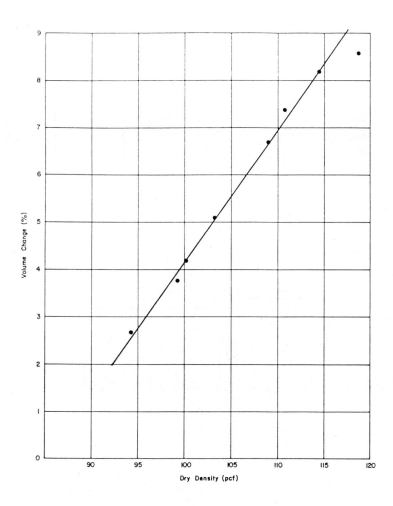

Figure 2-20. Effect of varying density on volume change for constant
              moisture content samples.

The dry density and swelling pressure relationship when plotted in a
rectangular scale is shown on Figure 2-22. The curve can be expressed by
the following exponential form:

$$y = kc^x$$
$$\text{where:} \quad k = 10^{-b}$$
$$c = 10^a$$

Table 2-7. *Effect of varying density on volume change and swelling pressure for constant moisture content samples.*

Initial density, pcf	Moisture content, percent		Initial degree of saturation, percent	Volume increase, percent	Swelling pressure, psf
	Initial	Final			
94.3	12.93	21.27	45.0	2.7	2,600
99.4	12.20	24.92	48.1	3.8	4,600
100.2	12.93	19.93	52.1	4.2	5,000
103.3	12.93	20.51	56.3	5.1	7,000
109.1	12.93	20.56	65.4	6.7	13,000
110.8	12.20	19.03	64.7	7.3	14,000
114.5	12.20	19.17	71.6	8.2	21,000
118.9	12.20	17.08	81.2	8.6	35,000
Average	12.55	21.08			

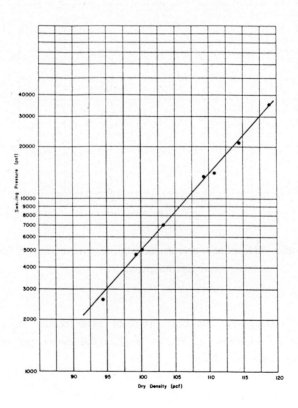

Figure 2-21.   Effect of varying density on swelling pressure for constant moisture content samples.

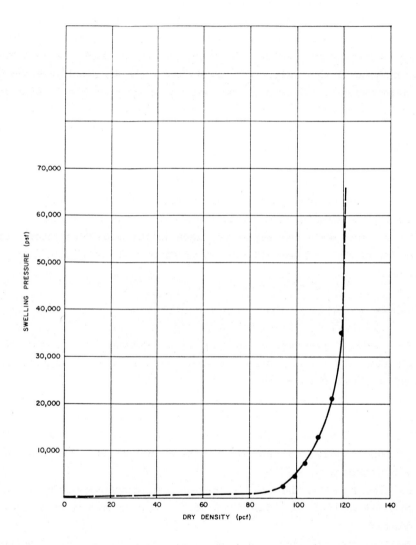

Figure 2-22. Effect of varying density on swelling pressure for constant
              moisture content samples.

When dry density decreases, swelling pressure rapidly approaches zero. When dry density increases, swelling pressure rapidly increases and approaches infinity. The soil engineer is interested, only within a narrow range of dry density, ranging from 100 to 130 pcf.

Since the foregoing established that the swelling pressure of a given soil is a constant and varies only with the dry density, swelling pressure can be conveniently used as a yardstick to measure the swelling characteristics of the soils. For undisturbed soil, dry density is the in situ characteristic. Therefore, the swelling pressure at the in situ dry density can be used directly to describe the swelling characteristics. For remolded soil, the swelling pressure varies with the degree of compaction. It is useful to introduce maximum Proctor density as a guide. In other words, the swelling pressure of remolded clay can be defined as the pressure required to keep the volume of a soil at its Proctor density constant.

In this particular case, the standard Proctor density is 108.4 pcf. At this density the swelling pressure is 12,000 psf. It is interesting to note that at the site where the sample was taken in its undisturbed state, the dry density of the soil was 110 pcf corresponding to a swelling pressure of 14,000 psf.

Conclusions

1. The swelling pressure of a clay is independent of the surcharge pressure, the initial moisture content, the degree of saturation, and the thickness of the stratum.
2. The swelling pressure increases with the increase of initial dry density.
3. For undisturbed soil, the swelling pressure can be defined as the pressure required to keep the volume of a soil at its natural dry density constant.
4. For remolded soil, the swelling pressure can be defined as the pressure required to keep the volume of a soil at its maximum Proctor density constant.
5. Swelling pressure can be used as a yardstick for measuring swelling soil. Swelling pressure reflects only the swelling characteristics of the soil and is not changed by placement conditions or environmental conditions.

LATERAL PRESSURE

In the design of earth-retaining structures, very little consideration
has been given to the lateral swelling pressure of the backfill soil.   In
the Rocky Mountain area, residential basement walls have been constructed
without earth pressure calculations.   It was generally assumed that the
backfill could be loosely compacted and sufficient support could be imposed
on the top and bottom of the wall to prevent excessive wall movement.
Basement walls were generally designed for an equivalent fluid pressure of
30 pcf to meet the requirements of the local building code.   No vertical
steel was called for in the design.   Since most basement walls are only
about seven feet high with a span of about 40 feet, failure of basement
walls was seldom reported.

Recently, however, in the highly expansive soil area in South Denver, a
number of houses have had bowing basement walls with horizontal deflection
as much as six inches reported.   In one case, the basement wall actually
toppled.   Figures 2-23 and 2-24 show the damaged basement wall under lateral
pressure.

Figure 2-23. Severe distortion of basement wall due to lateral
swelling pressure.

Figure 2-24. Total collapse of basement wall subjected to lateral
           swelling pressure.

The damage can be from several causes: the use of a large basement
escape window now required by a recently imposed fire code; inadequate wall
reinforcement; inadequate upper framing support; inadequate lower basement
slab support; and most important of all, the underestimation or the total
ignorance of lateral soil swelling pressure.

Pressure exerted on walls

Lateral earth pressure behind the basement walls can consist of the
conventional earth pressure, lateral swelling pressure exerted by backfill
against the wall, and the swelling pressure of natural soil against the
backfill. Figure 2-25 indicate the distribution of these pressures.
Conventional lateral earth pressure behind a basement wall depends on
the rigidity of the wall, the type of backfill, the degree of backfill
compaction and the surcharge load on the backfill. If the on-site soils
were pushed in without compaction, the density of backfill varies through
the section and depths. Without compaction effort, the lateral earth
pressure on the wall can be considered as an earth pressure at-rest

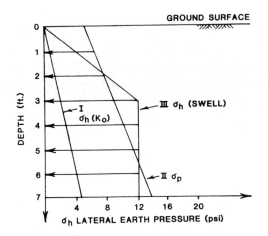

CASE I K_o Condition

        $C' = 2.1$ psi $\phi = 13.4°$ $Y = 109.5$ pcf

        $K_o = 1 - \sin \phi$ $\phi = 1 - \sin\ 13.4° = 0.89$

        At $Z = 0$ $\sigma_h = K_o\ \sigma_v = 0$

        $Z = 7$ $\sigma_h = K_o\ \sigma_v = 0.89 \times$
                       $109.5 \times 7 = 682$ psf $=$
                       4.7 psi

CASE II   Passive Earth Pressure

        $C' = 2.1$ psi $\phi = 13.4°$
        $Y = 109.5$ pcf

        $K_p = \tan^2 (45 + \frac{\phi}{2}) = 1.6$

        $\sigma_h = YZ\ K_p + 2C\sqrt{K_p}$

        At $Z = 0$ $\sigma_h = 2 \times 302.4 \times \sqrt{1.6}$
                       $= 765.8$ psf $= 5.3$ psi

        $Z = 7$ $\sigma_h = 109.5 \times 7 \times 1.6 + 2 \times$
                       $302.4 \times \sqrt{1.6}$
                       $= 1994.7$ psf $= 13.9$ psi

CASE III   Swell Pressure at Equilibrium State

        At $Z = 0$ $\sigma_h = 0$

        $Z = 3$ $\sigma_h = 12.0$ psi

        $Z = 7$ $\sigma_h = 12.0$ psi

Figure 2-25. Distribution of lateral earth pressure.

condition. In the earth pressure at-rest condition, the wall is assumed to
be rigid with no deflection. The pressure is expressed as

$$\sigma_h = \gamma Z K_o$$

Where K_o =   Coefficient of lateral earth pressure at rest
             γ =   Wet unit weight of soil mass
             Z =   Depth at which the lateral pressure is considered
             $\emptyset$ =   angle of internal friction

In this case under study, γ = 109.5 pcf, $\emptyset$ = 13.4° at depth seven feet, the
at-rest pressure is

$$\sigma_h = 109.5 \times 7 \times (1-\text{Sin } 6.6)$$
$$= 109.5 \times 7 \times 0.89$$
$$= 682 \text{ psf} = 4.7 \text{ psi}$$

If the backfill is assumed to be compacted and the wall is assumed to be
subject to deflection, then the distribution of lateral earth pressure can
be computed on the basis of an equivalent fluid unit weight. The
conventional approach, without testing and refined computation, is usually
assigned an arbitrary value of 35 to 50 pcf.

    Most of the basement backfill soils in expansive soil areas consist of
high plasticity clays apparently pushed in with no compaction. However, as
the backfill settles over a span of several years and as the soil undergoes
cycles of drying and wetting with the swelling pressure of the natural soil
pushed against the backfill, the density of the backfill can increase
considerably. Actual density tests performed in some of the residential
areas where the basements have caved in indicate that the backfill dry
density can reach 100 pcf. With such a density, the conventional earth
pressure can reach an equivalent fluid weight of more than 95 pcf.

Sample Description

    The sample used for testing was artificially prepared. The composite
clay sample was taken in the Denver area with some bentonite added. To
ensure uniformity of testing, the sample was pulverized to pass through a #8
sieve and the sample was cured in a moisture chamber to allow uniform
distribution of the moisture content.

The physical properties of the sample are as follows:

Liquid limit	= 107%
Plastic limit	= 25%
Plasticity index	= 82%
Passing #200 sieve	= 90%
Optimum moisture content	= 86.9 pcf

Using the conventional constant volume swell test, the laboratory results indicate a swelling pressure of 15.2 psi with an initial moisture content of 19.3%. Direct shear tests indicate that the sample has a $\emptyset=13.4°$ and a cohesion of 2.1 psi.

Lateral pressure testing setup

A simple small-scale testing setup was constructed to investigate the magnitude of lateral expansion pressure exerted on a rigid wall. The apparatus is shown on Figures 2-26 and 2-27. The walls surrounding the sample box were rigid so that the minimal deflection would take place during soil expansion. The sample was compacted into the box in one-inch layers at the maximum Proctor density. Filter paper was placed between layers to allow easy infiltration of water. The box measured 15"x5" in plan dimension and 12 inches in height. Initial moisture content was slightly lower than optimum. The sample was loaded at a 0.4 ksf initial surcharge and then controlled at a constant volume.

Three load cells were installed at the top, middle and bottom of the box to measure the lateral expansion pressure. An adjustable plate was placed at the top of the sample. Vertical expansion of the clay was controlled by a screw arrangement so that the sample was at constant height. Two load cells were installed at the top of the sample to measure the vertical swelling pressure. Any movement of the container walls could be detected by the strain gauges installed. Water was introduced into the soil through a steel mesh plate at the rear of the sample.

Test results on lateral expansion pressure are shown on Figures 2-28 and 2-29. It is seen that the lateral pressure increased steadily as the moisture infiltrated the soil and reached a peak value, then dropped as moisture increased. When complete saturation was attained, lateral pressure reached an equilibrium value of 12.0 psi.

Lateral swelling pressure

Extensive research on lateral swelling pressure has been conducted by R.K. Katti, Professor at Indian Institute of Technology [2-25] and Z. Ofer [2-26] at the University of Witwatersrand, Johannesburg, South Africa. Katti's tests under constant dead load indicate that "The lateral pressure increased rapidly with time in the beginning of the saturation process. Then the rate of increase slowed down and the lateral pressure attained a peak value. With further increase in time, the lateral pressure decreased to some extent and then remained constant." The result of constant volume

Figure 2-26.   Set-up for lateral expansion pressure measurement.

tests presented in Figure 2-28 is similar to the above finding.   It is interesting to note that at the peak value the ratio between upper lateral and vertical swelling pressure is on the order of 3.6.   At an equilibrium state, the ratio between vertical and lateral swelling pressure becomes unity.

The experiment was repeated using the same sample and with the same compaction procedure, but the filter papers between the compacted layers were eliminated.   As a result, seepage of water through the sample was very slow.   The experiment was carried on for 100 days.   At the peak value, the ratio between upper lateral and vertical swelling pressure is on the order

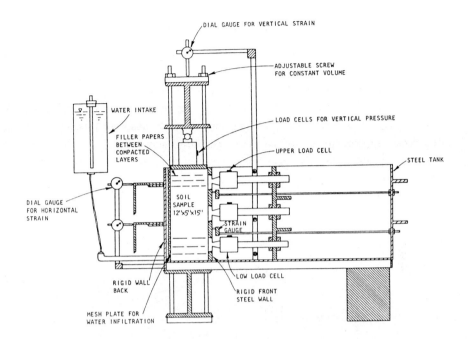

Figure 2-27. Lateral swelling pressure measurement set-up.

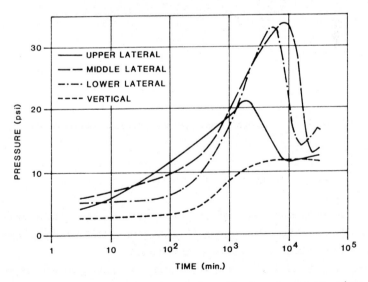

Figure 2-28. Lateral pressure and vertical pressure vs. time.

of 2.3.   After 100 days, lateral pressure approached vertical pressure as
shown on Figure 2-29.

The development of peak lateral expansion value can be interpreted from
the concept of clay particle arrangement.   Most of the expansive clay
mineral crystals have a sheet or plate arrangement [2-27] [2-28].   The
arrangement of the particles is flocculated or dispersed.   Flocculated soil
statistically has an isotropic expansion and dispersed soil has an
anisotropic expansion due to parallel particle arrangement.   The expansion
pressure perpendicular to the particle orientation is higher than the
expansion pressure in the parallel direction.

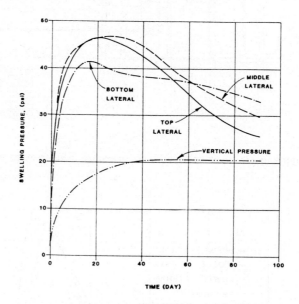

Figure 2-29.   Lateral pressure and vertical pressure vs. time.

Factors affecting lateral expansion pressure

It appears that the lateral expansion pressure is affected by the
initial dry density, the degree of saturation, the surcharge load and the
depth factor.   These are discussed as follows:

Dry Density: It is well known that vertical swelling increases with
the increase of dry density.   Z. Ofer [2-26] found that the swelling

pressure in both vertical and horizontal directions is related to initial density. The test was conducted under a constant vertical surcharge. The increase of lateral swelling pressure follows the same pattern as the increase of vertical swelling pressure with density.

Degree of Saturation: Katti [2-29] indicated that the lateral swelling pressure increases with water intake at first, then reaches a peak. After the peak, it decreases to an equilibrium state. This feature can also be expressed by lateral pressure against a time increment as water content increases with time. It is speculated that after the soil reaches saturation, an equilibrium condition is reached. The peak lateral swelling pressure always comes up before full saturation in both constant volume condition and constant vertical surcharge condition. From the tests it was observed that lateral swelling pressure is also influenced by the method used to saturate the soil. Water seeping horizontally into soils causes higher lateral pressure than vertical seeping. The filter papers in the sample allow water to enter into the sample vertically.

Surcharge Load: Both Katti [2-25] and Ofer [2-30] stressed the effect of the surcharge load on lateral expansion pressure. The equilibrium lateral pressure increases rapidly with the increase in initial surcharge. Under a high surcharge load, the increase in lateral pressure becomes less. The ratio of equilibrium lateral pressure to the vertical stress is as high as ten in the case of a low surcharge. The ratio decreases rapidly as the surcharge load increases.

Depth Factor: In a large-scale test, Katti [2-31] [2-32] found that lateral pressure at equilibrium changes from zero at the ground surface to a constant value at approximately 3.5 feet in depth. After this depth, a constant lateral pressure is maintained until vertical pressure due to the weight of the soil is equal to the swelling pressure. Katti also found that zero volume change occurs below approximately 3.5 feet. This test had no volume change below an equivalent depth.

Figure 2-30 shows the magnitude of lateral pressure at various heights and various stages. Stage one is the conventional condition of compacted soil with maximum pressure on the order of five psi. The second stage is the peak lateral expansion pressure. This pressure increases greatly with the increase of the conventional compaction pressure. However, when swelling pressure reaches equilibrium, no significant difference can be attributed to depth.

Conclusions

The small scale test the author conducted was to simulate the actual basement wall backfill condition in which the initial surcharge load (weight of soil) is small, about 2.5 psi, and the density is approaching maximum Proctor density. Under these conditions, the following conclusions can be reached:

1. Under K_o condition, on a rigid wall, the conventional earth pressure can reach 4.8 psi at the bottom of a seven-foot wall.

2. Lateral swelling pressure at a depth of approximately 3.5 feet can reach a peak value of 34 psi under a partially saturated condition. Such pressure is assumed to take place when poor surface drainage allows water to accumulate on the surface of the backfill. In other cases when a perched water table develops, water infiltrates into the backfill at the bottom of the wall. Thirty-four psi should be considered as the maximum possible force exerted on the wall. The actual pressure could only be 50% of the maximum value. In any case, the lateral swelling pressure exerted on the wall at a depth of about 3.5 feet below the surface far exceeds the conventional at-rest earth pressure. It is not difficult to realize that a poorly reinforced and poorly supported wall cannot withstand such pressure.

3. After several years when the backfill becomes fully saturated, an equilibrium lateral pressure on the order of 12 psi will be permanently exerted on the wall. Since the wall has already been damaged by the previous high peak swelling pressure, no further damage takes place. It is interesting to note that owners reported that wall damage took place about two years after occupancy with most damage taking place within a short span of time and no further movement occurring thereafter.

4. In the basic study of swelling pressure, the equilibrium lateral pressure should always be used. As presented in this test, the magnitude of lateral expansion pressure can vary with density, degree of saturation, surcharge load and depth consideration, but an equilibrium final value is reached. This value should be nearly equal to the vertical swelling pressure. The author believes that in the design of a permanent structure, such as a drilled pier foundation, equilibrium lateral expansion pressure should be used, but the possible effect of peak value should not be ignored.

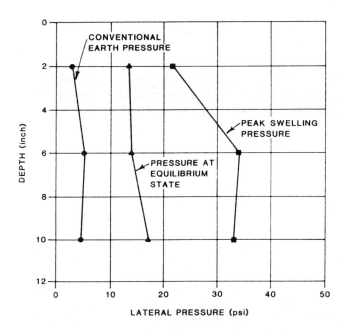

Figure 2-30. Lateral pressure distribution at different stages.

SHRINKAGE

   It is generally recognized that swelling and shrinkage of expansive
clays are interrelated, but it is uncertain if highly swelling clays also
show equally high shrinkage upon drying. It is doubtful that swelling can
be treated as an image reflection of shrinkage.

   Over a great portion of the world, shrinkage problems pose more of a
threat to structure damage than swelling problems. Notable clays are the
black cotton soil which covers much of India and a great part of Central
Africa. In China, where most of the expansive clays are of illite
mineralogy, shrinkage poses a much more severe problem than swell [2-33].
In Britain, problems associated with expansive soils are largely related to
the shrinkage caused by tree roots extracting soil moisture [2-34]. In the

Rocky Mountain region of the United States, shrinkage seldom results in
structural damage.   Continuous swelling appears to be the major concern.
The Taylor clays in San Antonio, Texas, the Yazoo clays in Jackson,
Mississippi, and the volcanic clays in Hawaii undergo seasonal movement.
Cracks open up during the dry seasons and close during wet months.

Soil properties and environmental conditions are the main factors
affecting the swelling and shrinkage behavior of expansive soils.   It is
believed that the same factors that affect swelling also control the
mechanics of shrinkage but at different degrees.

Effect of density on swelling and shrinkage

The typical clay used for the study was remolded clay shale from the
Denver area.   Its basic physical properties are as follows:

Liquid limit	48.5%
Plastic limit	22.4%
Specific gravity	2.83
Shrinkage limit	15.5%
Passing #200 sieve	93.1%

The clay shale was pulverized and prepared in the conventional manner
so as to obtain uniform test specimens for repeated testing.   The testing
was conducted in two series.   In the first series, dry density was held
constant and the moisture content was varied.   In the second series,
moisture content was held constant and the density was varied.

A conventional oedometer was used for testing.   The clay shale was
compacted into a ring 2.76 inches in diameter and 0.75 inch in height.   A
surcharge load of one psi was applied at the initial moisture content.   When
compression was completed, the sample was immersed in distilled water and
allowed to swell for a period of two to three days.   After swelling was
completed, water was drawn out without disturbing the sample and allowed to
air dry at room temperature.   At the same time, the weight of the sample was
recorded.   When the weight of the sample was reduced to the initial sample
weight, the moisture content and dry density were determined and recorded.
From the above data, the percentage of swelling and shrinkage was
determined.

The initial moisture content was 14.9% ± 0.25% and the dry density
varied from 90 pcf to 110 pcf.   Figure 2-31 is a plot of measured swelling

and shrinkage for various initial dry density. It is seen from Figure 2-31
that swelling increased rapidly with increase in initial dry density.
Shrinkage remained fairly uniform with an increase in initial density,
varying from 12% to 13%. It can be surmised that dry density variation has
little effect on shrinkage. In other words, if the initial moisture is kept
constant, a change in initial dry density does not substantially change the
shrinkage value. Considerable difficulty was experienced in controlling the
final moisture content when the sample was air dried from saturation to the
initial moisture content.

     It is tentatively concluded that there exists a point at which the
swelling equals the shrinkage. The initial dry density at this point is
defined as critical dry density. Critical dry density varies with soil
property and initial moisture content. If initial dry density is greater
than the critical dry density, swelling is larger than shrinkage. If
initial dry density is less than critical dry density, shrinkage is greater
than swelling.

Effect of moisture content on swelling and shrinkage

     Popescu [2-35] stated that there exists a critical moisture content
range within which shrinkage takes place. Beyond the critical moisture
content range, any further change of moisture content does not cause further
shrinkage. He further divided shrinkage into three stages: initial
shrinkage, normal shrinkage and residual shrinkage. In practice, variation
of moisture content in clay soil usually falls within the critical moisture
content range.

     In this study, the dry density was kept constant at 107.0 ± 0.6 pcf,
and initial moisture content varied from 15.1% to 22.3%. This range of
moisture content is believed to be within the critical moisture content
range as defined by Popescu. The test results are shown on Figure 2-32. It
is seen from Figure 2-32 that swelling decreases with the increase of
initial moisture content. It is postulated that shrinkage remains unchanged
and less than swelling when the initial moisture content is less than the
shrinkage limit. When the initial moisture content reaches the shrinkage
limit, shrinkage decrease with the increase of initial moisture content, and
the rate of decrease is lower than the rate of swelling. Therefore, with an
increase in initial moisture content, a point is reached at which the
shrinkage is equal to swelling, as shown on Point A in Figure 2-32. With
further increase in initial moisture content, the shrinkage is greater than

swelling. As further moisture increase continues, shrinkage goes from the
residual stage to the normal stage.

Within the normal stage, the rate of shrinkage decreases when the
increase in initial moisture content is at its highest level. This
shrinkage decreasing rate is larger than the rate of swelling decrease. At
Point B, shrinkage again is equal to swelling. Thereafter, swelling is
greater than shrinkage. As the initial moisture content approaches
saturation, the rate of both swelling and shrinkage decreases and finally
reaches zero.

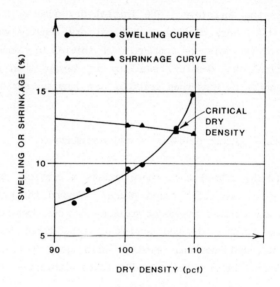

Figure 2-31. Effect of initial dry density on swelling and shrinkage.

Seasonal moisture fluctuation

It is believed that seasonal moisture content fluctuation can result in
heaving and settling of the structure. The cyclic up and down movement is
believed to occur in phases but lags behind rainfall [2-36]. Theoretically,
shrinkage can result in settlement, but there is very little evidence of
appreciable downward movement under covered areas in a building. Seasonal
fluctuation of moisture content along the edges of highway pavement or

parking areas can be expected, but at the central portion of a covered area, shrinkage seldom takes place even under a prolonged arid climate.

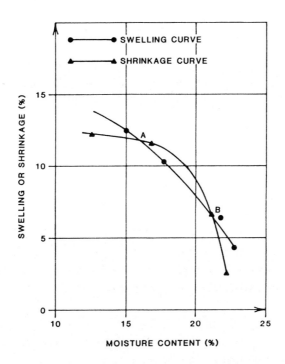

Figure 2-32. Effect of moisture content on swelling and shrinkage.

Movement measurements of buildings in good condition have seldom been performed. Monitoring cracked building movement cannot reflect the effect of seasonal moisture variation. An unusual opportunity afforded the making of a precise measurement of a school building founded on piers. The movement of the piers was measured each month for a period of 11 months. The movement of both the interior and exterior piers was recorded and plotted along with the average monthly precipitation, as shown in Figure 2-33.

The dry and wet periods coincide fairly well with the upward and downward movements of the exterior piers. However, for the interior piers, the movement graph lags behind the precipitation graph. This is expected

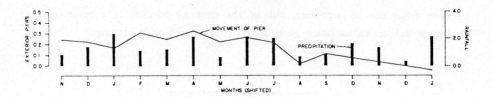

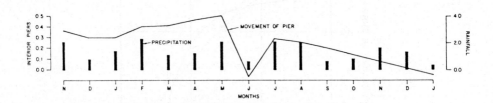

Figure 2-33. Movement of exterior and interior piers in a school building
              with respect to precipitation.

since it takes additional time for moisture or vapor to migrate from the
exterior of the building to the interior piers.

In the school building previously discussed, it should be noted that
over a period of 11 months, the difference between the maximum and minimum
pier movement seldom exceeded one-half inch. Such movement was insufficient
to manifest noticeable damage to the building. In fact, such movement would
not have been noted had not a precise leveling been conducted.

The end result of shrinkage around or beneath a covered area seldom
causes any structural damage and therefore is not an important item to be
considered by soil engineers. It is the continuous increase in moisture
content beneath the covered area that introduces the damage in an expansive
soil area.

Relationship between swelling and shrinkage

J. Uzan [2-37] measured swelling forces and shrinkage forces of
remolded heavy clay and concluded that the swelling force and the shrinkage
force are similar when compared on the basis of similar conditions. X.Q.

Chen and Z.W. Lu [2-38] claimed that swelling and shrinkage are reversible.

It is the author's opinion that shrinkage can be greater or smaller than swelling depending upon the initial state of the soil. It is questionable that swelling and shrinkage are reversible. The following points are noted:

1. Initial dry density has little effect on shrinkage but is a controlling factor on swelling.

2. There exists a critical dry density at which swelling is equal in magnitude to shrinkage. Critical dry density varies with soil property and initial moisture content. When initial dry density is greater than the critical dry density, swelling is greater than shrinkage. Shrinkage is greater than swelling with low initial dry density.

3. When dry density approaches the critical dry density, shrinkage and swelling are stabilized. Further wetting and drying do not affect the soil behavior.

4. Initial moisture content plays an important role in swelling and shrinkage relationship. Within the range between Points A and B in Figure 2-32, shrinkage is greater than swelling. Outside of this range, swelling is greater than shrinkage.

REFERENCES

[2-1]   "A Review Paper on Expansive Clay Soils," by Woodward-Clyde
        Associates, Vol. 1, 1967.

[2-2]   Seed, H.B., Woodward, R.J. and Lundgren, R., "Prediction of Swelling
        Potential for Compacted Clay," Journal ASCE, Soil Mechanics and
        Foundation Div., Vol. 88, 1962.

[2-3]   Lambe, T.W. and Whitman, R.V., "The Role of Effective Stress in the
        Behavior of Expansive Soils," Quarterly of the Colorado School of
        Mines, Vol. 54, No. 4, October, 1959.

[2-4]   Snethen, D.R., "Evaluation of Expedient Methods for Identification
        and Classification of Potentially Expansive Soils," 5th
        International Conference on Expansive Soils, South Australia.

[2-5]   "Special Procedures for Testing Soil and Rock for Engineering
        Purposes," 5th Edition, 1970, ASTM, STP 479, 1970.

[2-6]   Jennings, J.E. and Knight, K., "The Prediction of Total Heave from
        the Double Oedometer Test," 1957-58 Symposium on Expansive Clays,
        South African Institution of Civil Engineers, Johannesburg.

[2-7]   Design Manual, NAVFAC DM-71, Department of the Navy, Naval
        Facilities Engineering Command, 1982.

[2-8]   Van Der Merwe, D.H., "The Prediction of Heave from the Plasticity
        Index and Percentage Clay Fraction of Soils," The Civil Engineer in
        South Africa, Vol. 6, No. 6, 1964.

[2-9]   Fredlund, D.G., "Prediction of Ground Movements in Swelling Clays,
        31st Annual Soil Mechanics and Foundation Engineering Conference,
        University of Minnesota, 1983.

[2-10] Johnson and Snethen, "Prediction of Potential Heave of Swelling
        Soil," ASTM, 1979.

[2-11] McDowell, C., "The Relation of Laboratory Testing to Design for
        Pavements and Structures of Expansive Soils, "Quarterly of the
        Colorado School of Mines, Vol. 54, No. 4, 1959.

[2-12] Seed, H.B., Woodward, R.J., Jr., and Lundgren, R., "Prediction of
        Swelling Potential for Compacted Clays," Journal of the Soil
        Mechanics and Foundations Division; Proceedings of the American
        Society of Civil Engineers, Vol. 88, No. SM3, 1962.

[2-13] Vijayvergiya, V.N. and Ghazzaly, O.I., "Prediction of Swelling
        Potential for Natural Clays," in Proceedings of the Third
        International Conference on Expansive Clay Soils, Vol. 1, Jerusalem
        Academic Press, Jerusalem, 1973.

[2-14] Vijayvergiya, V.N. and Sullivan, R.A., "Simple Technique for Identifying Heave Potential," in Proceedings of the Workshop on Expansive Clays and Shales in Highway Design and Construction, Vol. 1, National Technical Information Service, Springfield, Va., 1973.

[2-15] Nayak, N.V. and Christensen, R.W., "Swelling Characteristics of Compacted Expansive Soils," Clays and Clay Minerals, Vol. 19, No. 4, 1974.

[2-16] Schneider, G.L. and Poor, A.R., "The Prediction of Soil Heave and Swell Pressures Developed by an Expansive Clay," Research Report TR-9-74, University of Texas Construction Research Center, Arlington, Tex., 1974.

[2-17] McKeen, R.G., "Characterizing Expansive Soils for Design," paper presented at the joint meeting of the Texas, New Mexico, and Mexico Sections of the American Society of Civil Engineers, Albuquerque, N. Mex., 1977.

[2-18] Johnson, L.D., "Predicting Potential Heave and Heave with Time in Swelling Foundation Soils," Technical Report S-78-7, U.S. Army Corps of Engineers Waterways Experiment Station, Vicksburg, Miss., 1978.

[2-19] Aitchison, G.D. and Richards, B.G., "The Fundamental Mechanics Involved in Heave and Soil Moisture Movement and the Engineering Properties of Soils which are Important in Such Movement," Second International Research and Engineering Conference on Expansive Clay Soils, Texas A & M Press, 1969.

[2-20] Wong, S.L., "Properties of Expansive Soil and Foundation Design," Chinese Academy of Building Research, 1980.

[2-21] M. Ali, El-Fatih and A.D. Elturabi, Muawia, "Comparison of Two Methods for Measurement of Swelling Pressure," 5th International Conference on Expansive Soils, 1984.

[2-22] Porter, Andrew A. and Nelson, John D., "Strain Controlled Testing of Expansive Soils," 4th ICOES, 1980.

[2-23] Kassiff, G. and Baker, R., "Aging Effects on Swell Potential of Compacted Clay," Journal of the Soil Mechanics and Foundation Division, ASCE, Vol. 97, SM 3 Proc. March, 1971.

[2-24] Kassiff, G. and Shalom, A.B., "Experimental Relationship Between Swell Pressure and Suction," Geotechnique, Vol. XXI, No. 3, September, 1971.

[2-25] Joshi, R.P. and Katti, R.K., "Lateral Pressure Development Under Surcharges," 1980, Proc. of 4th International Conf. on Expansive Soils, U.S.A.

108   FOUNDATIONS ON EXPANSIVE SOILS

[2-26] Ofer, Z., "Instruments for Laboratory and In Situ Measurement of the Lateral Swelling Pressure of Expansive Clays," 1980, Proc. of 4th International Conf. on Expansive Soils, U.S.A.

[2-27] Lambe, T. William, "Compacted Clay: Engineering Behavior," 1960, Transactions, ASCE, Paper 1655, Vol. 125, pp. 718-741.

[2-28] Lambe, T. William, "Compacted Clay: Structure," 1960, Transactions, ASCE, Paper 3041, Vol. 125, Part 1.

[2-29] Katti, R.K., Kulkarni, S.K. and Foledar, S.K., "Shear Strength and Swelling Pressure Characteristics of Expansive Soils," 1969, Proc. of 2nd International Conf. on Expansive Soils.

[2-30] Ofer, Z., "Laboratory Instrument for Measuring Lateral Sol Pressure and Swelling Pressure," Geotechnical Testing Journal GTJODJ, Vol. 4, No. 4, Dec. 1981.

[2-31] Katti, R.K., Lal, R.K., Foledar, S.K. and Kulkarni, S.K., "Depth Effects in Expansive Clays," 1969, 2nd International Conf. on Expansive Soils.

[2-32] Katti, R.K., Moza, K.K., and Kulkarni, U.V., "Shear Strength Development in Expansive Soil With and Without CNS Surcharge," Proc. of 5th International Conf. on Expansive Soils, Adelaide, South Australia, 1984.

[2-33] Chen, F.H., "The Distribution and Characteristics of Swelling Clays in China," Proc. of 5th International Conference on Expansive Soils, Adelaide, South Australia, 1984.

[2-34] Driscoll, R., "The Influence of Vegetation on Swelling and Shrinkage of Clay Soils in Britain." Geotechnique, Vol. 33, No. 2, 1983.

[2-35] Popescu, M., "Behavior of Expansive Soils with a Crumb Structure," 4th International Conference on Expansive Soils, 1980.

[2-36] Jennings, J.E., "The Theory and Practice of Construction on Partially Saturated Soils as Applied to South Africa Conditions." "Engineering Effects of Moisture Changes in Soils," International Research and Engineering Conference on Expansive Clay Soils, Texas A & M Press, 1965.

[2-37] Uzan, J., etc., "Two Dimensional Restrained Shrinkage of Remolded Heavy Clay," 3rd International Conference on Expansive Soils, 1973.

[2-38] Chen, X.Q. and Lu, Z.W., "Calculation of Movement of Building Foundation of Expansive Soils," 5th International Conference on Expansive Soils, 1984.

Chapter 3

FIELD AND LABORATORY INVESTIGATIONS

INTRODUCTION

The stability of a structure founded on expansive soil depends upon the subsoil conditions, ground surface features, type of construction, and possibly even the meteorological variations. Subsoil conditions can be explored by drilling and sampling, seismic surveying, excavating test pits, and by studying existing data. Ground surface features are controlled by surface geology and physiography. A study of existing structures in the immediate vicinity of the site can be of prime importance in determining the type of construction in an expansive soil area. This is necessary in the investigation of soil for a building addition.

Elaborate site investigation oftentimes cannot be conducted due to limited assigned construction costs. For very favorable sites, elaborate site investigation may not be warranted. However, if the area is suspected of having swelling soil problems, extensive soil investigation is necessary even for very minor structures. Generally, it is the small building with inadequate funding, insufficient planning, and a low-bidding contractor who unwisely economizes in constructing the building that presents the most problems. The soil engineer should not accept jobs in an expansive soil area which does not allow a thorough subsoil investigation.

SITE INVESTIGATION

Before initiating the site investigation, the soil engineer should obtain information regarding site topography, surficial geology, and existing structures. This can be accomplished by reviewing available data, studying topographic and geologic maps, and by making a reconnaissance survey.

Topography

The topographic condition is an essential part of the site investigation. Generally, for larger projects a topographic survey is available. However, care must be taken to ensure that the survey is correct.   Many times, site grading completely alters the survey.   The elevation of each test hole should be recorded.

One important aspect is the selection of a bench mark.   Every effort should be made to tie in the elevations with the architect's reference point.   Since the floor level governs the selection of foundation type, this aspect cannot be overemphasized.   Whenever possible, the bench mark should be referenced to establish data such as an existing building, manhole invert, cross on a sidewalk, top of fire hydrant and so forth.

The location of natural and man-made drainage features is also of considerable importance.   Erecting a structure across a natural gully always poses a future drainage problem.   The water level in any nearby streams and rivers should be measured and recorded.   Irrigation ditches carry large amounts of water during the irrigation season.   Water leaking from a ditch can supply moisture to the foundation soil and cause swelling of footings and slabs.   Pier uplift due to the infiltration of water from an irrigation ditch is not uncommon.   Water leaking from the ditches can also cause other problems, such as basement damage.   The location and elevation of the ditch should be included as a part of the field records.

Streams and nearby rivers naturally are of importance in the site investigation.   Of particular interest, in avoiding flooding, is the extent of flood plains.   Such preliminary information can usually be obtained from the U.S. Geological Survey and the U.S. Department of Agriculture soil survey reports.

The steepness of valley slopes is of special concern for sites to be located in mountainous areas.   Some environmental agents classify valley slopes in excess of 30 degrees as potential hazard areas.   Slope stability depends upon the slope angle of the rock and soil formation, evidence of past slope movement, and drainage features.   The field engineer should be aware of the possible slope problems associated with landslides, local slope failure, mudflow or other problems.   The vegetative cover on the slope, the shape of the trees, and the behavior of any neighboring structures should also be observed.

Much emphasis has been placed on the topographical features as a factor affecting the damage incurred on structures founded on expansive soils [3-1].   The Chinese engineers claimed that the extent of damage for structures founded on sloping ground is more severe than those on flat ground.   By flat ground, they refer to ground surface with less than a five percent slope.   Also, structures placed on top of the slope or middle of the slope are less favorable than structures placed on the toe of the slope.

Researches indicate that for structures founded on flat ground, the swelling movement continues even in dry years. The amount of heave increases each year. Foundation movement tends to match seasonal moisture variation.

Experiences in China indicate that the most important factor affecting the subsoil moisture content is the topographical feature [3-2]. When natural slopes exceed five percent, the supply of ground water is not uniform, but the surface drainage is good. The fluctuation of ground water has a much greater effect than surface water or the climate condition. Consequently, the site is not considered to be favorable. On the other hand, when the natural slope is less than five percent, the supply of ground water is uniform but the drainage is generally poor; when the fluctuation of ground water is less than twice the depth affected by surface drainage, the site is considered favorable.

Surficial geology

General surficial geology of the area includes the study of slopes, tributary valleys, landslides, springs and seeps, sinkholes, exposed rock sections, and the nature of the unconsolidated overburden.

An inspection of upland and valley slopes may provide clues to the thickness and sequence of formations and rock structure. The shape and character of channels and the nature of the soil (residual, colluvial, alluvial, etc.) provide evidence of the past geologic activity.

An engineering geologist should identify and describe all geologic formations visible at the surface and note their topographic positions. The local dip and strike of the formations should be determined and note made of any stratigraphic relationships or structural features that may cause problems of seepage, excessive water loss, or sliding of the embankment.

To ensure adequate planned development of a subdivision, Senate Bill 35 of Colorado requires the subdivider submit items such as:
1. Reports concerning streams, lakes, topography, geology, soils, and vegetation.
2. Reports concerning geologic characteristics of the area that would significantly affect the land use and the determination of the impact of such characteristics on the proposed subdivision.
3. Maps and tables concerning suitability of types of soil in the proposed subdivision.

Similar bills exist in many other states in response to the environmental zeal of the nation. In mountain areas, the mapping of surficial geologic features is highly desirable. Johnson [3-3] suggested that the features to be shown on the map should include:

1. Texture of surficial deposits,

2. Structure of bedrock, including dip and strike, faults or features, stratification, porosity and permeability, schistosity, and weathered zones,

3. Ground-water features, including seeps, springs, observable water table, and drainage,

4. Area of modern deposit (result of accelerated erosion).

5. Unstable slopes, slips, and landslides, and

6. Faults and fault zones.

Of special importance is that the geologist should be able to recognize a potential swelling soil problem. For instance, the red siltstone formation in Laramie, Wyoming does not pose a swelling problem while a few miles to the east where the claystone of the Pierre Formation is encountered, swelling is critical.

Existing structures

The behavior of existing structures has an important bearing on the selection of the proposed structure. All possible information should be obtained concerning structures in the immediate proximity. Inquiry should be made as to the condition of the structure, age, and type of the foundation. If adjacent existing structures have experienced water problems, the possibility of a high-water table condition in the area exists. If the existing structure exhibits cracks, this usually indicates that the existing foundation system is not adequate. An experienced field engineer can readily identify the type of distress which has occurred in the existing buildings and then determine if it is associated with swelling soils. The age of the building should also be considered. If the cracks are old and new cracks have not appeared in the patched areas, the foundation movement may have been stabilized. However, even if the existing structure is in excellent condition, this does not mean that the existing foundation system should be duplicated in new construction. The existing structure could have been originally overdesigned. This is especially true for old structures where massive foundation systems were used.

DRILLING AND SAMPLING

Geophysical techniques are generally used for preliminary subsoil investigation. Both the seismic and the resistivity geophysical methods have been successfully used. Geophysical equipment has been commonly used in determining the surface of bedrock, major changes in subsoil conditions, and ground-water depth. However, such a survey should always be supplemented by drilling.

Probably, the most accurate subsoil investigation method is by the opening of test pits. In a test pit, the field engineer can examine in detail the subsoil strata, stratification, layers and lenses, as well as taking samples at the desired location. However, the depth of the test pit is limited to the reach of a backhoe, generally 12 feet. Also, when the water table is high, test pit investigation becomes useless. In locations where the subsoils consist essentially of large boulders and cobbles, the use of test pit investigation is most favorable. Auger drilling through boulders and cobbles is difficult, if not impossible, and in addition the cost of rotary drilling may not be warranted for small projects.

Another possible investigation method is to drill a large-diameter caisson hole to the required depth. By physically entering the caisson hole, the subsoil strata can be clearly examined, and undisturbed samples can be obtained at the desired depth.

The drawback of both the test pit method and the caisson hole method is that standard penetration tests cannot be performed. Probably 95 percent of all site investigations are conducted by drilling test holes, either by auger drilling, rotary drilling, percussion drilling, or other methods.

Test holes

The number of test holes required for a project depends upon the type of foundation system, uniformity of subsoil conditions, and to some extent the importance of the structure.

If preliminary investigation indicates that shallow spread footings are the most likely type of foundation, then it will be desirable to drill shallow test holes close together to better evaluate the subsoil conditions within the loaded depth of the footings. If, however, the shallow foundation system is not feasible and a deep foundation system is likely to be required, then the number of test holes can be decreased and the spacing increased.

As a rule of thumb, test holes should be spaced at a distance of 50 to 100 feet. In no case should the test holes be spaced more than 100 feet apart in an expansive soil area.

The client sometimes has the misconception that drilling of test holes is the major cost of the subsoil investigation, and consequently, there is a tendency to drill as few holes as possible, preferably only one. The risk involved in such an undertaking is enormous. Erratic subsoil conditions can exist between widely spaced test holes; therefore, logs of these test holes cannot be considered as representative of the overall site subsoil condition.

Closely spaced test holes are especially important where the presence of expansive soils is suspected. For instance, if sandstone and siltstone bedrock are encountered at a shallow depth, a logical recommendation is to found the structure directly on the bedrock with spread footings designed for high pressure. However, if more test holes are drilled, it is quite possible that claystone bedrock containing high swelling potential is revealed. The presence of swelling soil, even in only one test hole, can change the foundation recommendations completely.

The depth of test holes required is generally governed by the type of foundation. Before drilling, the type of foundation cannot be completely envisioned. Therefore, the first hole the field engineer drills should be a hole deep enough to provide information pertinent to both shallow and deep foundation systems. Samples in the hole should be taken at frequent vertical intervals, preferably not more than five feet apart. After the completion of the deep test hole, the field engineer should have a fairly good idea of the possible foundation system. Consequently, for the subsequent holes, more attention should be directed to the upper soil if a shallow foundation is likely. If, however, a deep foundation system is contemplated, drilling to bedrock is necessary for all holes.

Oftentimes, the depth to bedrock is a criterion as to the depth of test holes. Where bedrock is within economical reach, say, within 40 feet, it is advisable to drill a few holes into the bedrock. In Colorado, the depth to the top of shale bedrock is extremely erratic, and the depth to bedrock can increase as much as 30 feet within a short distance of 100 feet. This should be taken into consideration when determining the location and depth of test holes.

In some cases, deep holes are required, not from the subsoil requirement but for the determination of the water table elevation. For deep basement construction, the depth of test holes should be at least 20 feet to preclude the possibility of ground water becoming a problem in the lower floor.

Penetration test

In performing the standard penetration test, a soil sampler known as a split spoon is used. It is an open-ended steel cylinder which is split longitudinally into halves. These two halves are held together by a cutting show at the lower end and a coupling which connects the sampler to the drill rod. The split spoon is driven 18 inches into the ground by means of a 140-pound hammer falling from a free height of 30 inches. The number of hammer blows required to drive 12 inches is called the standard penetration resistance N which represents the number of blows per foot, or the blow count.

The standard penetration test has been and will continue to be an important and practical field test. The following should be considered in using the penetration test:

1. The value of N in cohesionless soils is influenced by the depth at which the test is made. At a great depth, the same soil with the same relative density would give a high penetration resistance. The influence of overburden pressure can be approximated by the following equation: [3-4]

$$N = N'\left(\frac{50}{p+10}\right)$$

where   N = Adjusted value of standard penetration resistance,

   N' = Standard penetration resistance as actually recorded, and

   p = Effective overburden pressure, psi

The author found that while the increase in penetration resistance at a greater depth cannot be ignored, the calculated values are rather high. Using the above equation and a value of N' = 12, at 20 feet below the ground surface with γ = 110 pcf, the adjusted penetration resistance is almost doubled. This results in the designing engineer using a penetration resistance value larger than required for a safe design.

2. Penetration resistance is reliable only if the driving condition is not abused. The hammer should be entirely free falling without being subject to undue friction. The standard penetration barrel should not

be packed by overdriving since this forces the soil against the sides of the barrel and causes incorrect readings. An increase in blow count by as much as 50 percent can sometimes be caused by a packed barrel. Driving a standard barrel into gravelly soils presents a problem. The barrel bounces when driving on cobbles, and hence, no useful value can be obtained. Sometimes, a small piece of gravel jams in the barrel thereby preventing the entrance of soil into the barrel, thus substantially increasing the blow count.

3.   Considerable economy can be achieved by combining penetration resistance with sampling. A slight modification in the design of the barrel will allow the insertion of thin wall lining in the barrel and provide a blow count as well as an undisturbed sample. The modified barrel is commonly referred to as the California sampler. Field tests have been conducted comparing the results of the penetration resistance of the California sampler with that of the standard penetration tests. The tests indicate that the results are commensurable with the exception of very soft soil (N <4) and very stiff or dense soil (N >30). By combining penetration resistance tests with the sampling device, more tests can be made and undisturbed samples can be obtained without resorting to the use of Shelby tubes.

Sampling

Some contracts call for a penetration test every five feet and sampling for the same interval or every change of soil stratum. This may prove unnecessary. The field engineer should use his judgment to guide the frequency of sampling and avoid unnecessary sampling so that the cost of investigation can be held to a minimum. Samples in the upper 10 to 15 feet are important, as this is generally the site of shallow foundations; also, soil characteristics at this level govern the slab-on-ground construction and earth-retaining structures. Sampling and penetration tests at lower depth, say, in bedrock, become critical when a deep foundation system is required.

Sampling above the proposed floor level should be limited since this material will be excavated. Sometimes, for deep basement construction, no sampling or penetration tests are necessary for the first 15 feet below ground level.

Sampling and testing at an intermediate depth generally are not too critical, except when friction piers are under consideration. Instead of assigning the field engineer definite instructions as to the frequency of sampling, it would be better to leave the matter up to his discretion.

A reasonably good sample can be obtained when driving into shale bedrock. Auger drilling in most cases can be successfully conducted in shale bedrock. For other types of bedrock, such as limestone and granite, rotary drilling is necessary and rock cores obtained. Core samples are brought up by the drill and can be visually examined. The general characteristics, particularly the percentage of recovery, are of importance to foundation design and construction.

LABORATORY TESTING

Soil testing is essential in establishing the design criteria. Distinction must be made between the needs of the practicing engineer and those of the research engineer. For a practicing engineer, the purpose of laboratory testing is mainly to confirm his preconceived concept derived from field drilling, penetration tests, visual examination of the samples, and personal experience in the area. Exotic laboratory equipment and refined analysis are in the realm of the research engineer. Neither time nor budget will allow the practicing engineer to follow the researcher's procedure.

Swell test

The most important laboratory test on expansive soils is the swell test. The standard one-dimensional consolidation test apparatus is similar to that used in most soil laboratories for consolidation studies. This apparatus, known as a consolidometer, is shown on Figure 3-1 and is of the fixed-ring type [3-5]. It can accommodate a remolded or undisturbed sample from 2 to 4.25 inches in diameter and from 0.75 to 1.25 inches in thickness. Porous stones are provided at each end of the specimen for drainage or saturation. The sample unit is placed on a platform scale table and the load is applied by a yoke actuated by a screw jack. The load imposed on the specimen is measured by the scale beam and a dial gauge is provided to measure the vertical movement.

The advantage of such apparatus is that it is possible to hold the upper loading bar at a constant volume and allow the measurement of the maximum uplift pressure of the soil without volume change. This requires a constant load adjustment by an operator. A more advanced scheme is to use the consolidometer with a triaxial frame along with the attachments and electrical circuits [3-6]. Such a device allows automatic load increment and measures swelling pressure without allowing volume change to take place.

The consolidometer can also be used to measure the amount of expansion under various loading conditions. Since swelling pressure can be evaluated by loading the swelled sample to its original volume as explained in Chapter 2, it is simple to convert the platform scale consolidometer into a lever-type consolidometer as shown on Figure 3-2. A simple consolidometer can be made locally with low cost. The average soil laboratory should have a train of consolidometers to speed up testing procedures.

Interpretation of test results

Laboratory testing of disturbed and undisturbed soils can be performed in most soil laboratories by an efficient laboratory technician. Laboratory test results are reliable only to the extent of the condition of the sample. Results of testing on badly disturbed samples or samples not representative of the strata not only are useless but also add confusion to the complete analysis.

Equally important to testing of representative samples is the frequency of testing. Testing of a few samples on a single project and basing the final analysis on such testing is not only undesirable but sometimes dangerous, especially when evaluating swelling soils. The potential of swelling generally cannot be determined visually, and laboratory swell tests must be conducted. By testing only a few samples, the high swelling sample may have been missed and erroneous conclusions drawn. Too little testing is worse than no testing at all.

Interpretation of test results should be conducted by an experienced engineer, having the ability to screen the test results and exclude the dubious ones, to determine when the minimum or maximum values are to be used, and when average values should be used.

Judgment

Researchers of swelling soil usually assume that the soil is homogeneous or that it has defined layers with distinct physical properties; they use such assumptions to advance various theories predicting the behavior of structures founded on such soils. All practicing engineers realize that homogeneous soils seldom exist, especially in the case of

Figure 3-1. Platform scale consolidometer

Figure 3-2. Simplified lever-type consolidometer

swelling soil. The swelling potential varies drastically with depth and
with extent. It is not uncommon that one house in a subdivision is
underlain by nonexpansive soils and the neighboring houses have serious
uplifting problems.

Drill holes are never placed closely enough and sampling is never taken
at too frequent intervals. Since economic considerations dictate the amount
of field investigation a client can afford, much will depend on the good
judgment of the geotechnical engineer.

For the advancement of the science of expansive soil, it is deemed
necessary to research the very complicated swelling soil behavior. We have
to know the effect of soil disturbance during sampling, we have to know the
effect of preconsolidation, and we have to know the limitations of
laboratory testing. Theories must be advanced so that mathematical formulas
can be used to express soil behavior.

Yet, to the practicing engineer, all the refined analysis and
calculation will be upset by taking samples 10 feet below the representative
sample or by moving the drill hole 20 feet away from the representative hole
location. The following questions always arise by the geotechnical
engineers engaged in practical design and recommendations:

1. Shall I take the highest swelling value in all the testing as design
    standard, or shall I take the average value?
2. Shall I assume that the environment will not be altered after
    construction? Would perched water develop at a later stage, thus
    upsetting my design criteria?
3. Is the drill log reliable? Are there lenses of hard rock or soft soil
    which the field engineer neglected to log?

It may be advisable that the geotechnical engineer take a course on
"theory of probability," so that the data obtained can be intelligently
interpreted. It is possible that such knowledge can be more valuable than
all the refined mathematical analysis on soil characteristics.

Ralph Peck in his article, "Where has all Judgment Gone," [3-7] stated,
"When an engineer at my age talks about judgment, he invites the criticism
that he is too old to keep up with the latest advances in theory and methods
of calculation and so, having slipped behind the times he must depend on a
somewhat vague attribute called judgment." This statement is appropriate
and realistic in dealing with expansive soils. No amount of sampling and
testing and analysis can replace judgment.

If the area where the proposed structure is to be located has a history
of problem heaving, no matter what the investigation results are, the

engineer must approach the design with extreme caution. On the other hand, it is always possible that the soil will exhibit swelling potential when subjected to drying and subsequent wetting. If the environment is such that excessive drying is unlikely, as in the case of the Hawaiian Islands, then the chance of severe damage from expansion will be remote.

REFERENCES

[3-1]   Chen, F.H., "The Distribution Characteristics of Expansive Soils in China," 5th International Conference on Expansive Soils, Adelaide, Australia, 1984.

[3-2]   Wong, S.L. and Wang, W.M., "Structure Deformation on Expansive Soils During Drought Years," Chinese Academy of Building Research, 1980.

[3-3]   Johnson, A.I., "suggested Method for Geologic Reconnaissance of Construction Site," Special Procedures for Testing Soil and Rock for Engineering Purposes," ASTM stp 479, 1970.

[3-4]   Teng, Wayne C., "Foundation Design," Prentice-Hall, Inc., 1962.

[3-5]   Lambe, T.W., "Soil Testing for Engineers," John Wiley & Son, Inc.

[3-6]   Agarwal, K.P., and Sharma, S.C., "A Method for Measuring Swelling Pressure of an Expansive Soil," Proceedings of the Third International Conference on Expansive Soils, Haifa, Israel.

[3-7]   Peck, Ralph R., "Where has all Judgment Gone," National Research Council of Canada, 1960.

Chapter 4

DRILLED PIER FOUNDATIONS

INTRODUCTION

Drilled pier, drilled caisson, or caisson foundation is widely used in the Rocky Mountain area. Drilled piers, when made with an enlarged base, are commonly referred to as belled piers and when made without an enlarged base are referred to as straight-shaft piers.

The drilled pier foundation is used to transfer the structural load from the upper unstable soil to the lower stable soil. The use of a drilled pier foundation covers a wide range of possibilities, some of which follow:

1. Piers drilled into hard bedrock for supporting high column load,
2. Friction piers bottomed on stiff clays for supporting light structures,
3. Belled piers bottomed on sand and gravel for supporting medium column load, and
4. Long, small-diameter piers drilled into a zone unaffected by moisture change in swelling soil areas.

The drilled pier foundation is a rational solution to combat the problem of expansive soils; however, the design and construction must be closely controlled.

As stated by Richard Woodward [4-1], "If investigation and design were always perfect; supervision always competent and adequate, inspection always continuous, experienced and alert; and the contractor's personnel always expert and conscientious—if all these conditions prevailed all the time on a project, then there would be no defective piers."

PIER CAPACITY

Piers bearing on bedrock or shale have been designed using mostly empirical considerations derived from experience, from limited load test data, and from the behavior of the existing structures. For instance, because of its erratic characteristics, the ultimate load carrying capacity of Denver Blue Shale has never been determined. Nonetheless, piers supporting a column load in excess of 1,000 kips are commonly designed and constructed in the Rocky Mountain area with satisfactory performance.

Woodward, Gardener and Greer [4-1] commented on the empirical design of piers: "Many piers, particularly where rock bearing is used, have been designed using strictly empirical considerations which are derived from regional experience." They further stated that "Where subsurface conditions are well established and are relatively uniform, and the performance of past constructions well documented, the design by experience approach is usually found to be satisfactory."

The bearing capacity of drilled piers bottomed on bedrock is a combination of the end-bearing capacity and the skin friction developed between the pier wall and bedrock.

Bearing capacity

The bearing capacity of bedrock for piles or piers can be found in most building codes. The values given are generally conservative and the type of bedrock usually not clearly defined.   The methods for determining the bearing capacity are as follows:

1.   Penetration resistance of bedrock,
2.   Unconfined compressive strength or triaxial shear strength of undisturbed bedrock core sample,
3.   Consolidation tests on bedrock core sample under high load,
4.   Actual record of settlement of existing building founded on similar bedrock,
5.   Load test on bedrock, and
6.   Menard pressuremeter test on bedrock in the test holes.

Unfortunately, all of the above approaches have their limitations. Penetration resistance on hard bedrock involves blow counts in excess of 100.   When a soft seam or a very hard lens is encountered, the actual penetration resistance of the stratum cannot be accurately determined. Therefore, to obtain an accurate determination of penetration resistance, many tests, possibly more than 50, should be conducted.   After the representative penetration resistance value has been determined, the soil engineer should convert the blow count to allowable bearing capacity.   One widely used method is:

$$q_a = \frac{N}{2}$$

where: q_a = allowable bearing capacity in ksf, and

               N = blow count

For instance, if the representative blow count is 40, the allowable bearing capacity selected will be 20,000 psf. This approach is quite conservative. Extensive laboratory testing indicates that a more realistic value for q_a should be in the range of N to 0.75N. That is, with N = 40, the allowable bearing capacity should be about 30,000 to 40,000 psf.

Allowable bearing capacity derived by penetration resistance data should be checked by both the unconfined compressive strength test and the consolidation test. Unconfined compressive strength test at best can only represent the lower limit of the actual bearing capacity of bedrock. Unconfined compressive strength performed on drive samples can only be used for comparative purposes. Samples obtained from large diameter cores obtained from core drilling are far more reliable, but the presence of slickensides and other effects of disturbance limit the validity of such tests.

A high capacity consolidation test on a reasonably good sample can sometimes be used to determine the amount of pier settlement. The actual settlement value taken as a percentage of laboratory consolidation value should be left to the judgment of the soil engineer.

Probably the most reliable method of estimating bedrock-bearing capacity is the observation of the behavior of existing structures. Unfortunately, such records are scarce. In the Denver area, piers bearing on the Denver Blue Shale Formation designed for a bearing capacity of 60,000 psf have performed satisfactorily with long-term settlement of less than one inch.

Load tests on piers are costly and time consuming, especially for large-diameter piers. Limited data on the testing of small-diameter piers indicates that the conventional approaches are extremely conservative.

The most direct and recommended method in determining bearing capacity is the insertion of a Menard pressuremeter into the bottom of the test hole and evaluating the actual bearing capacity. An experienced operator is required to conduct the tests.

Research conducted by Jubenville and Hepworth [4-2] stated the following:

"The design of low capacity drilled piers in the Denver Metropolitan area is based on limited data collection and an empirical relationship which is very dependent upon results of the standard penetration test. The relationship should be used with caution and understanding; however, it appears that based on the excellent performance of innumerable structures designed with this procedure, it can be used successfully by the practicing engineer."

Skin friction of shale

The load-carrying capacity of a pier depends not only upon its end-bearing value but also to a great extent on the skin friction or side shear between the concrete and its surrounding soils. The skin friction value is unfortunately difficult to determine as discussed later under "Friction Piers."

Soil engineers in the Rocky Mountain area chose the establishment of skin friction value by the following two assumptions:

1. Assuming that skin friction value in bedrock is one-tenth of the end bearing value of bedrock, then for an end bearing value of 60,000 psf, the skin friction value is 6,000 psf. This value obviously has exceeded the cohesion of claystone and the total horizontal pressure against the shaft surface. It is generally thought that the skin friction value may not develop fully along the surface of the shaft but that a lubricated surface may exist between the soil and the shaft. Consequently, attempts are made to provide a shear ring in the portion of the pier in bedrock. Shear rings are made by grooving tools which cut slots around the circumference of the hole. The circular slots are about one-inch deep and 3/4-inch thick at intervals of eight inches. It is believed that by grooving the hole, adequate development of skin friction can be assured. Experience shows that such an installation is seldom justified. The walls of a pier hole are seldom smooth, with the exceptions of those drilled into oily shale. If it is found that a pier hole is too smooth, a tooth protruding about one-half inch can be inserted onto an auger and with several turns the surface of the hole can thus be roughened artificially.

    Load tests on small diameter piers (12 inches) drilled into bedrock with no end bearing (pier poured on elastic material) indicate that the skin friction value exceeds the design value by several times. It is believed that for small diameter piers, skin friction has actually taken all the load exerted on the pier without transferring the load to the pier bottom.

2. In designing the pier load, the skin friction developed in the portion of pier embedded in the overburden soil is usually ignored. Such values are considered to be added factors of safety in the pier

system. Actually, for long piers, the skin friction value developed in the overburden soils can be considerable. Assuming the drilling of a 42-inch-diameter pier through 20 feet of dense sand and gravel into the bedrock, the unit skin friction value between the concrete pier and the granular soil is at least 400 psf. Total skin friction value can exceed 90 kips. This added factor of safety to the drilled pier system is recommended. For major structures, it is important that an adequate factor of safety be incorporated to allow for unforeseen construction contingencies.

With the development of a perched water condition, the role of skin friction in the mechanics of a drilled pier becomes very complex. The following can take place:

1. Perched water travels through the seams and fissures in shale with some reaching the interface between the pier and pier shaft. The skin friction can be partially or completely lost and swelling along the pier shaft takes over.

2. After several years of subsoil under a perched water condition, it is entirely possible that the bearing value as well as the skin friction value can decrease. This is evident in the case of a drastic decrease in penetration resistance before and several years after the building is completed.

It is also possible that due to imperfect pier installation techniques, the design skin friction does not develop and there can be disturbed material present along the pier shaft.

Design capacity

After end bearing and skin friction values have been established, it is not difficult to calculate the total load-bearing capacity of a pier. Experience with shale bedrock in the Rocky Mountain area generally indicates that the hardness of bedrock increases with depth. This has been verified by penetration resistance and by pressuremeter tests. Therefore, for piers penetrating deep into bedrock, a higher load-carrying capacity can be assigned. Experience indicates that both the end-bearing capacity and the skin friction value increase with depth at the rate of three percent per foot. Using this assumption, the load-carrying capacity of a straight-shaft pier can be expressed as follows:

$$Q = A(E + 0.03EL)+0.1(E + 0.03EL)CL$$

in which: Q = total pier carrying capacity (kips)

                 A = end area of pier (sq. ft.)

                 E = end bearing capacity of pier (psf)

                 L = depth of penetration into bedrock (ft.)

                 C = perimeter of pier (ft.)

Figure 4-1 was prepared using the above equation, with the assumption that the end bearing value of top of bedrock is 50,000 psf. It should be pointed out that the above equation has its limitations as noted:

1. The bearing capacity of the top of bedrock should be carefully evaluated. Oftentimes, the top of the shale is highly weathered and only a low value can be assigned. Sometimes, deep excavation will expose the top of shale and the material will be subject to severe disintegration. Therefore, it is usually safe to assign a bearing-capacity value for that portion of the pier two to four feet below the surface of bedrock and then increase the bearing value for additional penetration.

2. The increase of bearing capacity with the depth of penetration has its limitations. According to the formula, the bearing capacity of the pier will be doubled when the depth of penetration exceeds 30 feet. Obviously, a load-carrying capacity of such magnitude cannot be assigned to the pier. Extensive experience in the Rocky Mountain area indicates that the absolute maximum of end-bearing value for piers in shale is 100,000 psf or about 700 psi.

Using the above approach in designing drilled piers, the load-carrying capacity of a single 60-inch-diameter pier can easily reach 3,000 kips with reasonable penetration. Such a load is of sufficient magnitude to accommodate most column loads of a high-rise structure.

MECHANICS OF PIER UPLIFT

As stated earlier in the chapter, the principle of drilled piers is to provide a relatively inexpensive way of transferring the structural loads down to stable material or to a stable zone where moisture changes are improbable. There should be no direct contact between the soil and the

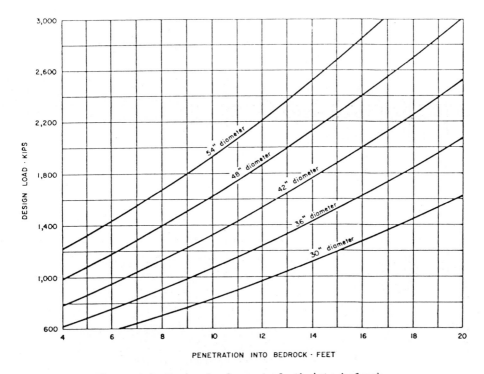

Figure 4-1 Design load versus depth into bedrock

structure with the exception of the soils supporting the piers. Figure 4-2 is a sketch of the grade beam and pier system.

Figure 4-2 shows that the uplifting forces which tend to pull the pier out of the ground are a direct function of the swelling pressure. The withholding force consists of the dead-load pressure exerted on the pier and the skin friction along the unwetted portion of the pier. For safe pier design, the withholding force must balance the uplifting force. These forces are analyzed as follows.

Uplifting force

The total uplifting force of the soils surrounding the pier can be written as follows:

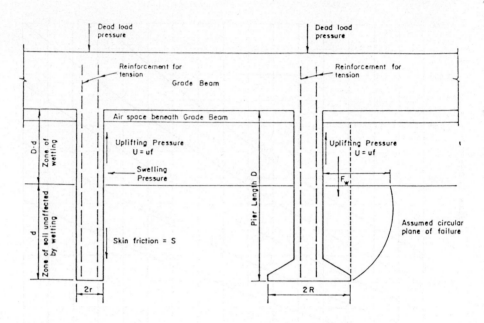

STRAIGHT SHAFT-PIER FOUNDATION   BELL-PIER FOUNDATION

Figure 4-2 Grade beam and pier system

$$U = 2 \pi \, rfu \, (D-d)$$

where: r = radius of the pier (ft.)

d = depth of the zone of soils unaffected by
wetting (ft.)

D = total length of the pier, (ft.)

u = swelling pressure, (psf)

f = coefficient of uplift between concrete and soil,

U = total uplifting force, (lbs.)

The soils surrounding the pier expand both vertically and horizontally
upon wetting. Parcher and Liu [4-3] have shown that compacted clay soils
exhibit greater unit swelling in the horizontal direction than in the
vertical direction. Since the magnitude of the difference of swelling is

small, the vertical swelling pressure can be used in estimating the
uplifting force on the piers.

The coefficient of uplift between the soil and surface of the concrete
pier is not known. With the model pier test described in the subsequent
section, an attempt was made to establish a rational value for the
coefficient of uplift for design purposes. It should be pointed out that
the uplifting pressure of the pier may not be controlled by the shear
strength of the clay. Our experiment, discussed later under "Model Test for
Pier Uplift," indicates that the pier actually slips out of the soil in very
much the same manner as extracting a pile. The contact surface between the
concrete and the soil remains clean.

Withholding force

The withholding force that keeps a pier from pulling out of the ground
can be written as:

$$W = \pi r^2 p + 2\pi r \, sd$$

where: p = unit dead-load pressure, (psf)

          s = skin friction surrounding the pier, (psf)

          W = total withholding force, (lbs.)

          d = depth of zone of soils unaffected by wetting, (ft.)

          r = radius of the pier, (ft.)

The skin friction of the soils surrounding the pier is again a factor
which cannot be fully evaluated. According to Mohan and Chandra [4-4], "The
frictional resistance of bored concrete piles in medium-to-hard clays is
about half the average undisturbed shear strength of the clay along the pile
shaft and has the same value in loading and pulling." Experience with piers
embedded in shale indicates that the skin friction along the shaft of the
pier is equal to at least 1/10 of the end-bearing value of the pier. The
value is conservative and depends upon the roughness of the pier shaft and
the compressive strength of the concrete, as previously discussed under
"Skin Friction of Shale."

Zone of wetting

In pier design it is important to determine the depth of the zone of wetting. Field tests indicate that expansive soils are so impermeable that surface water can penetrate only about two feet into the soil. However, in actual cases, it was found that soils surrounding piers were wet to a depth of as much as fifteen feet below the ground surface. This is possibly due to the following reasons:

1. The source of wetting may not be derived from surface water. A broken water main or sewer pipe can cause a much more severe wetting condition than surface water.
2. Surface water seeps through seams and fractures in the stiff fissured clays and clay shales.
3. Water tends to seep along the walls of the pier shaft into the surrounding soils.

For the above reasons, it is not possible to precisely define the depth of the zone of wetting. For design purposes, five feet can be arbitrarily assigned as the probable depth of wetting.

Model test for pier uplift forces

The soil sample used for conducting the test was typical of southeast Denver's highly expansive clay. The physical properties of the soil were as follows:

Liquid limit, percent	= 50.0
Plastic limit, percent	= 24.3
Plasticity index, percent	= 25.7
Passing No. 200 sieve, percent	= 82.4
Optimum moisture content, percent	= 21.0
Maximum dry density, pcf	= 96.5

The soil was packed into a steel container to a depth of nine inches in thin layers and at a moisture content of 14.0 percent. Two sets of holes were drilled in the soil.

The first set of two piers was prepared for studying the behavior of friction piers. The piers were two inches in diameter and were bottomed on the steel container. The length of the pier was embedded in the full depth of soil which was nine inches; therefore, any uplift pressure exerted on the pier was only along the pier shaft and none on the base.

The second set of two piers was prepared for studying the behavior of end-bearing piers. Holes 2-1/2 inches in diameter and six inches deep were drilled in the soil sample. The piers, being only two inches in diameter, did not have any direct contact with the soil along their circumference but were bearing directly on the soil at the bottom of the hole. Consequently, the only pressure acting on these piers was at the base.

The arrangement of the tests is shown on Figures 4-3 and 4-4. Piers A and B are friction piers. The uplifting movement of Pier A was recorded by placing a dial gauge directly on the top of the pier. The uplifting pressure of Pier B was measured with a proving ring which was centered on top of the pier with a ball bearing between the pier and proving ring. An adjustable screw was placed beneath the ball bearing. Dial A, which was placed on top of the pier, was carefully observed for movement; as soon as uplifting movement was observed on Dial A, the adjustable screw was turned to keep the pier in its original position throughout the test. The stress registered by the proving ring represents the true uplifting pressure of the pier [4-5]. Piers C and D were end-bearing piers. The set up for these two piers was identical to Piers A and B.

Figure 4-3 Apparatus for the determination of the coefficient of uplift

Perforated plastic pipes were inserted around the periphery of the piers as shown on Figure 4-5. Water was added to the soil through the perforated pipes to obtain saturated conditions. Both stress and strain

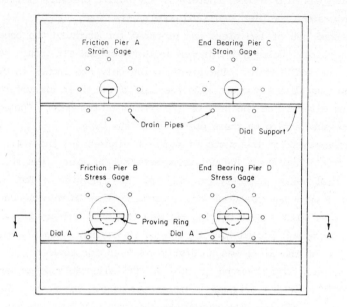

PIER UPLIFTING TEST APPARATUS
PLAN VIEW

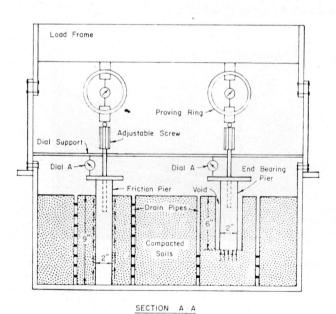

SECTION A·A

Figure 4-4 Pier uplifting test apparatus--plan and section

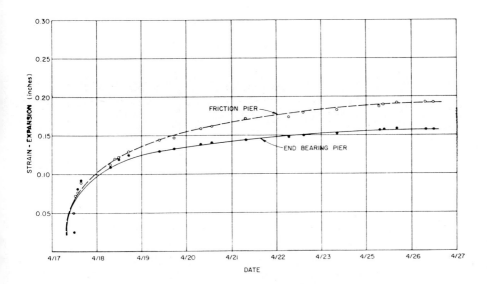

Figure 4-5 Time-strain relationship of friction and end-bearing piers in
            clay

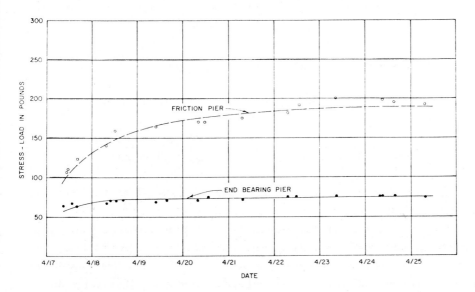

Figure 4-6 Time-stress relationship of friction and end-bearing piers in
            clay

readings were taken at frequent intervals. The test was carried out for a
period of two weeks. Results of the time-strain and time-stress
relationships are shown on Figures 4-5 and 4-6 respectively.

The results of the tests can be summarized as follows:

Pier No.	Type of pier	Maximum uplifting pressure, pounds	Maximum upward movement, inches
A	Friction	--	0.20
B	Friction	195.0	--
C	End Bearing	--	0.16
D	End Bearing	77.0	--

For complete wetting conditions and without pressure being exerted on the
bottom of the piers, the total uplifting force is:

$$U = 2\pi r f u D$$

The total uplifting force exerted on an end-bearing pier with the pier
free from the surrounding soils can be written as:

$$U = \pi r^2 u$$

Substituting the actual uplifting pressure measured in the model test;

For end-bearing piers: For friction piers:

$$77 = \pi(1)^2 u$$ $$195 = 2\pi(1) fu(9)$$
$$u = 24.5 \text{ psi}$$ $$fu = 3.45 \text{ psi}$$

Assuming that the vertical swelling pressure is equal to the horizontal
swelling pressure, then fu = 3.45 psi, from which the value f is 0.14.

From the results of the experiment, and from experience with pier
systems in this area, it appears that the uplifting pressure along the
surface of the concrete exerted by the soil in soil-concrete pier systems is
about 15 percent of the vertical swelling pressure. This value is of
significant magnitude and can be the governing factor in the design of the

grade beam and pier systems. The following example will illustrate the importance of uplifting pressure along the surface of the pier.

Assume a 12-inch-diameter pier embedded ten feet in expansive soils having a swelling pressure of 10,000 psf. Assuming that the upper five feet of the soil becomes wetted, then the total uplifting force exerted on the pier is:

$$U = 2\pi r f u (D-d)$$
$$= 2\pi(0.5)(0.15)(10,000)(10-5)$$
$$= 23,600 \text{ lbs.}$$

For piers spaced on about 10-to-15-foot centers, the dead-load pressure normally assigned to the piers is about 20,000 psf. The total dead-load pressure exerted on a 12-inch-diameter pier is 15,700 lbs. The unbalanced uplifting pressure is 23,600-15,700 = 7,900 lbs. Compensation must be made for this unbalanced uplifting pressure. This is accomplished by the pier in the lower unwetted portion of the soil where friction along the surface of the pier provides a restraint to uplift.

In extreme cases, the entire length of soil surrounding the pier is wetted. The total uplifting pressure is 47,200 lbs. and the unbalanced uplifting pressure is 31,500 lbs. Under this condition, uplifting of the pier is inevitable and structural cracking results.

It is interesting to know the shape of the failure plane when the pier is lifted from the ground. One theory is that when the pier is being lifted, a cone of soil is carried by the pier as shown on Figure 4-7a. By convention, the cone is assumed to be 60°. Another theory is that the pier is pulled out from the ground relative to the surrounding soils as shown on Figure 4-7b. In our model test, we found that after the completion of the test, Pier A had actually pulled from the surrounding soils leaving a gap between the bottom of the pier and the surface of the container, a distance approximately equal to the vertical movement of the pier, as indicated on the strain gauge. This indicated that the failure took place along the interface of soil and concrete as indicated on Figure 4-7b.

It was concluded from the above model pier tests that the uplifting pressure along the surface of concrete was about fifteen percent of the vertical swelling pressure as previously discussed. Theoretical analysis of the magnitude of the uplifting pressure was approached by Michael O'Neill [4-6] by the application of the concept of effective stresses against pier shaft during swelling. It is assumed that the coefficient of uplift between

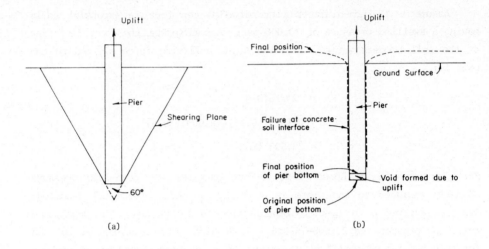

Figure 4-7 Soil failure plane resulting from pier uplift

concrete and soil is a function of the residual angle of the internal
friction of the soil. The magnitude of swelling pressure (u) should also be
qualified. Vertical swelling pressure is normally obtained in an oedometer
in which zero lateral strain is permitted, making u larger than would occur
in the field where lateral strain is not zero. O'Neill suggested to modify
the value u with an empirical factor, probably greater than 1.0.

Strictly, the value u should be the lateral swelling pressure of the
soil instead of the vertical swelling pressure. As explained in Chapter 2,
at the initial stage of wetting, lateral swelling pressure can be several
times greater than vertical swelling pressure. As the moisture migration
pattern stabilizes, vertical and horizontal swelling pressure tend to
equalize. Such behavior can be attributed to the phenomenon that pier
uplift usually takes place shortly after pier installation and becomes
stabilized after several years.

Assuming that the coefficient of uplift is equal to the residual angle
of internal friction, the theoretical total uplift can be estimated. The
residual angle of internal friction for stiff clay and clay shale is on the
order of 5 to 10 degrees [4-7]. This would give an uplift coefficient on
the order of 0.09 to 0.18 or 9 to 18% of the swelling pressure. The 15%
value assumed by the author appears to be reasonable.

Rational pier formula

A rational pier formula can be derived by equating the total uplifting force and the total withholding force acting on a pier.

$$2\,\pi fu(D-d) = \pi r^2 p + 2\pi rsd$$

$$\text{then:}\quad p = \frac{2}{r}\,[fu(D-d) - sd]$$

$$\text{where}\quad p = \text{unit dead load pressure}$$

To solve the above equation, it is necessary to assign values to u, f, s, and D-d. The expansive soils usually encountered in this area belong to the categories of medium or high degree of expansion (Table 1-4). For the usual design purposes, it is possible to assign values of the soil properties in the above equation and obtain rational solutions. The following assumptions are made to simplify the computation:

1. The soil is uniform for the full length of the pier. If shale bedrock is encountered at the bottom of the pier, the error will be on the safe side.
2. Surface wetting affects only the upper five feet of the pier.
3. Skin friction, s, of the soils surrounding the pier in the unwetted zone is about 500 psf.
4. The swelling pressure acting on the pier for soils with a high degree of expansion is about 10,000 psf and for soils with a medium degree of expansion it is about 5,000 psf.
5. The coefficient of uplift, f, between the pier and the soil is 0.15.

Based on the above simplified assumptions, Figure 4-8 was prepared. From the figure, it is possible for the structural engineer to select the size of the pier, the dead-load pressure, and the required total length of the pier. Figure 4-9 shows an effective, drilled pier foundation system.

BELLED PIERS

Piers drilled into materials other than bedrock are often enlarged at the bottom of the hole for the purpose of increasing the bearing area, thus increasing the total load-carrying capacity. Such piers are commonly referred to as belled piers, or underreamed piers. The ideal bell is in the shape of a frustum with a vertical side at the bottom. The vertical side is possibly six to twelve inches high. The sloping side of the bell should be

at an angle of at least 60 degrees with the horizontal. Most drillers are
capable of providing bells with diameters equal to three times the diameter
of the shaft.

The advantages and disadvantages of the belled system are presented in
the following two paragraphs.

Advantage of belled piers

In Figure 4-2, the uplift forces exerted on the belled pier system are
as follows:

$$U = P + F_w + F_s$$

where: U = total uplifting force due to the swelling of
                 the soils surrounding the pier shaft in the
                 unwetted zone, (lbs.)

           P = total vertical pressure exerted on the pier, (lbs.)

           F_s = total shearing resistance along the assumed
                 circular line of failure in the unwetted zone,
                 diameter of the top of frustum = 2R, (lbs.)

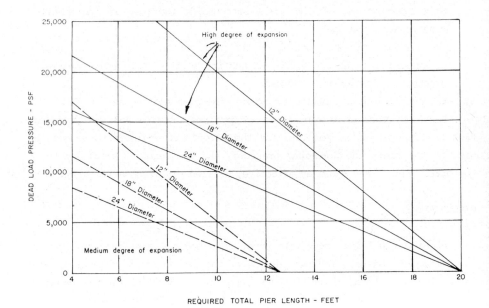

Figure 4-8 Rational pier design chart

Figure 4-9 An effective drilled pier foundation system

Probably the greatest advantage of the belled pier is that the
resistance against uplift is not affected by loss of friction in the zone
unaffected by wetting.   As previously discussed under "Pier Uplift," the
withholding force depends upon the skin friction of the pier in the zone
unaffected by wetting.   If, for some reason, such as the rise of ground
water, the skin friction is lost, then pier uplift is unavoidable.   With
belled piers, the total weight of the soil above the bell, F_w, is not
affected by moisture change.   Hence, there is always an added factor of
safety of this system against uplift.

In areas where the upper soils are highly expansive, the bedrock is
shallow, and there is a strong possibility of the development of a perched
water table condition, the use of a belled pier system should be favorably
considered.

Since a belled pier system relies entirely upon the anchorage of the
lower portion of the pier in the zone unaffected by wetting against uplift,
this type of pier can be used for columns with very light loads.   The
magnitude of dead-load pressure exerted on the pier is not a factor.

A most effective situation in which belled piers can be used
successfully is where the upper expansive soils are underlain by
nonexpansive, granular soils.   An oversized pier hole can be drilled to the

top of the granular stratum. A small diameter Sonotube can then be inserted into the hole and the annular spaces filled with sawdust. Lastly, the Sonotube can be filled with concrete, as shown on Fig. 4-10.

In this case, there is no uplifting pressure exerted on the pier shaft. Water may enter through the void space around the pier into the bearing soil, but since the bearing soil consists of nonexpansive sand and gravel, no harm is done.

This is an ideal situation in which belled piers in combination with sleeved pier shaft can be used successfully. However, such a subsoil condition is not frequently encountered.

Disadvantage of belled piers

The shaft of a belled pier must be sufficiently large to allow cleanout and inspection. The minimum shaft size is 24 inches, although 30 inches is desirable. For straight-shaft piers with a large shaft diameter, the uplift pressure exerted on the pier is several times higher than a smaller 12-inch-diameter shaft. More reinforcement is required in a belled pier than in a small-diameter straight-shaft pier. If the reinforcement is inadequate, tension cracks can develop in the vicinity of the pier shaft and the bell.

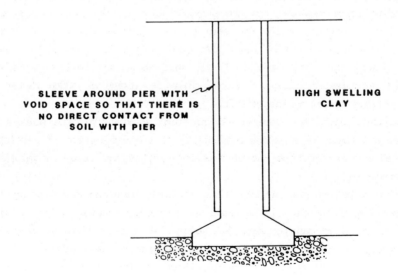

SLEEVE AROUND PIER WITH
VOID SPACE SO THAT THERE IS
NO DIRECT CONTACT FROM
SOIL WITH PIER

HIGH SWELLING
CLAY

Figure 4-10 Bell pier founded on sand and gravel

The most prominent disadvantages of belled piers are the cost and the difficulty of inspection. Bells can be formed mechanically with a special belling device. In hard bedrock, a mechanical device may be unsuccessful. When it is necessary to send a man down the hole with a jackhammer to complete the bell, the cost can be prohibitive. In the Denver area where the cost of pier drilling is exceedingly competitive, it is estimated that in a favorable drilling condition, the cost of belling can easily exceed the cost of drilling an extra ten feet of pier shaft. The advantage of anchorage in a bell pier system can easily be offset by drilling an additional ten feet into bedrock with a straight-shaft pier.

Cleaning a belled pier is much more difficult than cleaning a straight-shaft pier. Proper cleaning of a belled pier can only be accomplished by sending a man down the hole with a shovel. With present safety regulations, this means that all pier shafts have to be cased. The condition of the bottom of the straight-shaft pier can be inspected from the ground level with a torch or a mirror, but the condition of the bell cannot be inspected from ground level.

The difficulty of belled pier construction increases markedly where the materials immediately overlying bedrock are subject to caving. It is nearly impossible to bell in granular soil. The problem is further complicated if ground water is encountered.

With due consideration of the cost, the advantages of the belled pier can be achieved by the use of straight-shaft piers with a larger penetration into a zone unaffected by wetting.

Isolation of pier shaft

Since the most damaging action in a drilled pier system is the uplifting pressure exerted on the perimeter of the pier, the simple solution appears to be isolating the pier shaft from the uplifting forces.

The obvious method is to drill an oversized pier hole and provide soft, compressible material around the pier so that the surrounding soils cannot exert uplifting pressure on the wall of the pier. However, by so doing, the total load carrying capacity of the pier is greatly reduced. Since most piers rely heavily on skin friction to carry the column load, elimination of the skin friction necessitates the use of large bells for support to justify the use of such a system. The shortcomings of this system are as follows:

1. The load carrying capacity of the pier is reduced to that of a pad footing foundation,

2. No lateral resistance is available for the system,

3. The annular space must be filled with compressible material such as sawdust, ash, or straw. The use of loose sand or pea gravel will be satisfactory initially, but repeated seasonal cycles of expansion and shrinkage pack the filling and allow the surrounding soils to grip the pier shaft, and

4. The annular space filled with compressible material affords a free path for water to reach the bottom of the pier and heave the entire pier from the bottom. Consequently, the purpose of placing the pier in a zone unaffected by wetting is defeated. The end result of using such isolated-shaft belled piers is no different from using individual pads. This shortcoming can be partially corrected by sealing the upper three feet of the annular space with compacted clay. The process is costly and in most cases not effective in preventing the seepage of water.

Another approach has been devised to break the bond between the pier and the clay. The concept was initiated in San Antonio, Texas [4-1] where highly expansive clays are found. A pipe is introduced into the pier. The concrete-filled pipe is designed to carry the column load at the top. The outside of the pipe from the top down to the bottom of the potentially wetted zone is coated with a bituminous mastic material. When the pier is lifted by swelling soils, the annulus of concrete outside the coated section breaks in tension, the mastic coating breaks the bond, and the upward force is limited by the viscosity of the mastic. Figure 4-11 illustrates this system. This concept has been experimented in Colorado recently with only limited success.

FRICTION PIERS

Since the purpose of a drilled pier foundation is to transfer the building load to a zone where the moisture content is not affected by surface wetting, it is, therefore, unnecessary to drill all piers into bedrock or very hard formations. Where the bedrock is deep and the upper overburden clays are expansive, a logical solution is the use of friction piers. In the design of a friction pier, a rational value should be assigned to the skin friction.

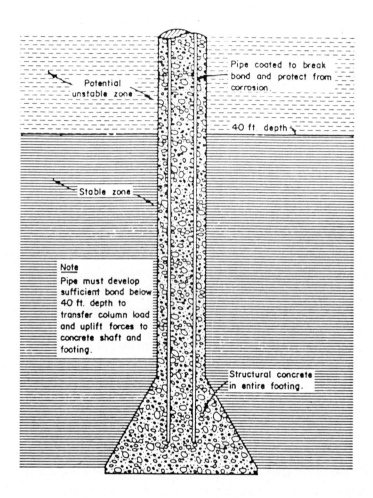

Figure 4-11   Design of belled pier for the relief of uplift due to
              expansion of upper clay layer. Note that the outer annulus
              of concrete is expected to break in tension near the bottom
              of the expansive clay layer. (Raba and Associates,
              Consulting Engineers, San Antonio, Tex.)

Skin friction

It is generally recognized that the skin friction between cohesive soils and the pier shaft cannot exceed the cohesion of the soil. Cohesion may be assumed to be equal to one-half of the unconfined compressive strength of the soil. In stiff or hard clays, however, the bond between the concrete and the soil may be less than cohesion. According to Teng [4-8], the maximum design value for skin friction of all cohesive soils should be limited to about 1,000 (or up to 1,500) psf.

Ultimate pier shaft resistance can be calculated as the sum of the ultimate shearing resistance imposed by the various strata which are in contact with the pier. The ultimate shaft resistance of a pier drilled in uniform clay strata can be expressed in terms of the undrained shear strength S_u and a reduction factor a as follows:

$$S = \pi r^2 \ aS_u$$

The undrained shear strength of clays can be determined in the laboratory from the unconsolidated-undrained triaxial test. For stiff clays, an unconfined compression test will generally serve the purpose. Other more direct methods in evaluating shear strength, such as the use of a pressuremeter inserted into the drill hole at the desired depth, can be used. A vane shear test is not applicable for use in stiff clays but can be valuable in soft clays. Shear value can also be interpreted indirectly from field penetration resistance tests or cone penetration resistance data.

The shear strength reduction factors as derived from analysis of prototype field load tests in clays, shales, and tills, are listed in Table 4-1.

The shear strength reduction factor, according to Woodward, Gardner and Greer [4-1], is also influenced by construction effects and the moisture sensitivity of the supporting materials. Water migration from the concrete to the pier wall also reduces the shear strength. Other elements, such as the duration of exposure of open shafts, can also be important factors.

Design of friction piers

The principle for the design of friction piers is given in a previous section under "Rational Pier Formula." The following considerations should be given:

Table 4-1. Shear strength reduction factors (a) for drilled piers

| Material | Material Properties | | | a | Reference |
	Moisture content W_n, percent	Plasticity Index I_p, percent	Shear strength, tons per sq. ft.		
Stiff clay	23	35-55	1.2	0.44	Whitaker and Cooke
Stiff clay	25	20-60	1.2	0.62	Reese and O'Neill
Massive shale	15	7-16	5.0	0.64*	Matich and Koziki
Glacial till	12	2-16	2.5	0.64*+	Matich and Kozicki
Stiff clay	---	----	1.1	0.52	Woodward et al.
			0.9	0.49	
Stiff clay	19	36-46	1.4	0.30	Mohan and Jain

*Failure was not reached.
+Sandy gravel with cobbles and approximately 50 percent silty clay, $N \geqslant 45$ blows per ft.
(After Woodward, Gardner, and Greer)

1. Friction piers generally bear on stiff clays. With the use of small diameter piers, the end-bearing capacity of such piers can generally be neglected.
2. Sufficient field penetration resistance tests should be performed not only to establish the proper friction value but also to ensure that soft soils are not encountered within the length of the pier.
3. Since the upper five feet of soil around the pier is subject to surface wetting and uplifting, this length should be excluded in calculating the pier load capacity.
4. Friction piers should not be used at a site where ground water is high or where there is the possibility of a future high ground-water condition.

An example of typical friction pier design is as follows:

Data   Stiff clay to depth 40 feet, no ground
                water encountered. Average penetration
                resistance   = 25 blows per foot
                Average unconfined compressive
                strength   = 4,000 psf
                Pier length   = 20 feet
                Pier diameter   = 12 inches
                Shear strength reduction factor   = 0.5

Pier capacity	Ultimate skin friction	$= \dfrac{4,000}{2}$ x 0.5
		= 1,000 psf
	Total load carrying capacity	
	= (20-5) x 1,000 x 3.14	= 47.1 kips
	Using a factor of safety of 3	
	Design load	= 15.7 kips
	When designing for a dead load plus a full live load, a factor of safety of 2 can be used.	
	Then design load	= 23.6 kips
Uplifting force	Swelling pressure	= 10,000 psf
	Coefficient of uplift	= 0.15
	Total uplift = 3.14 x 10,000 x	
	0.15 x 5	= 23.5 kips
Withholding Force	Ultimate skin friction	= 1,000 psf
	Ultimate withholding force	
	=(20-5) x 1,000 x 3.14	= 47.1 kips

The above calculations indicate that the design is safe as the withholding force is larger than the uplifting force.

FAILURE OF THE PIER SYSTEM

A properly designed drilled pier system involves the coordination of the floor slab, grade beam, void space, reinforcement, expansion joint, and floor system. A typical detail is shown on Figure 4-12.

The grade beam and pier system offers the most logical solution for lightly loaded structures founded on expansive soils. However, if incorrectly designed, or incorrectly constructed, a building with a pier foundation is just as vulnerable, if not more vulnerable, to movement than a building founded on spread footings.

Considerable experience is required to determine the cause of cracking of a building founded on piers. Oftentimes, the cracks are caused by slab movement as discussed in Chapter 6. Typical pier uplift movement generally takes place a short distance from the pier and has a 45-degree pattern.

Cracks are often wider at the top and narrower at the bottom. Generally, the type of cracking depends upon the structural configuration of the building.

Masonry walls and cinderblock walls are most sensitive to movement. Consequently, the first sign of pier movement is reflected in cracks that develop in the brick wall as shown on Figures 4-13 and 4-14. Concrete walls are structurally more resilient to differential movement than masonry walls. When severe diagonal cracks appear in the basement as shown on Figure 4-15, pier uplifting can be considered a certainty.

The most common errors in design are insufficient pier length, excessive pier diameter, or the absence of pier reinforcement. The most common errors in construction are excess concrete on top of the piers resulting in mushrooms at the top of the piers and the absence of or defective air space beneath the grade beams. These are discussed in detail as follows:

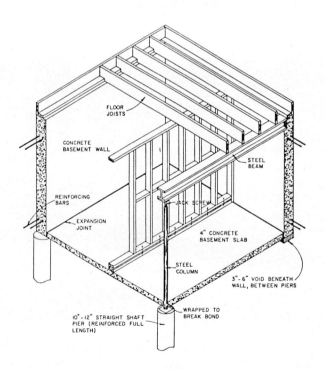

Figure 4-12   Typical detail of the grade beam and pier system

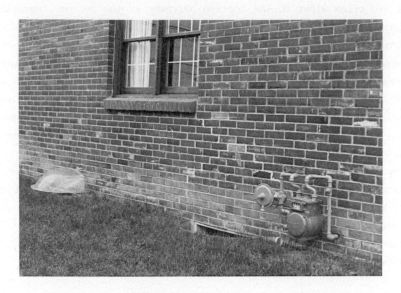

Figure 4-13   Typical cracking caused by pier uplift

Figure 4-14   Heaving of piers of a three-story structure.   Insufficient
pier length is the cause.

Excessive pier size

Many laymen have the impression that the larger the diameter of the pier, the safer the building. Actually, to exert enough dead-load pressure on the pier, it is necessary to use small diameter piers in combination with long spans. The most economical spacing of the piers is limited by the amount of reinforcement in the grade beams or the economical size of the floor beams. Normally, piers should have a minimum spacing of twelve feet.

Most small drill rigs in the Rocky Mountain area are equipped to drill 12-inch-diameter pier holes. Auger sizes of eight and ten inches are also available. However, with pier holes less than twelve inches in diameter, considerable difficulty can be encountered in cleaning the holes. A pier hole with as little as two inches of loose soil at the bottom can experience excessive settlement at a later date.

Figure 4-15 Typical cracks developing in the basement wall immediately
             beneath the window well. Note that the crack is wide at the
             top and narrow at the bottom.

Insufficient pier length

    As explained in the previous section, the stability of the pier against
uplift depends upon the amount of dead-load pressure exerted on the pier and
the anchorage provided in the lower portion of the pier. If the length of
the pier is short, the possibility of the soil at the bottom of the pier
becoming wetted is great; thus the skin friction along the pier providing
the anchorage is lost and the pier has to depend upon dead-load pressure
alone to resist uplifting.

    The dead-load pressure for a lightly loaded building is not of
sufficient magnitude to resist the uplifting. This is especially true in
the case of the interior, lightly loaded piers. Therefore, the function of
a short pier actually is no different or more desirable than individual pad
footings. This is illustrated on Figure 4-16.

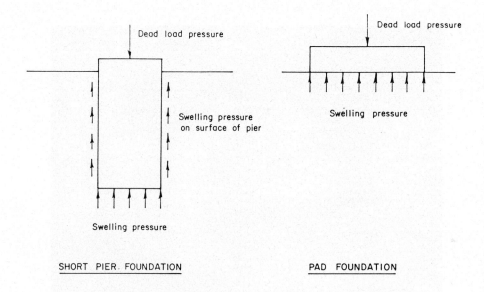

Figure 4-16 Swelling pressure acting on short pier foundation and pad
             foundation

    For short piers, the uplifting pressure is the sum of the swelling
pressure acting on the bottom of the pier plus the uplifting pressure acting
upon the perimeter of the pier.

Failure of the pier system due to insufficient pier length (Figure
4-17) is commonplace. The practice of using piers of insufficient length is
usually observed in areas where claystone shale is near the ground surface
and the engineer specifies only the depth of penetration into bedrock. For
example, it is common in this area to specify a minimum bedrock penetration
of four feet. In many instances, this results in a pier with a total length
of only four feet. There is a good possibility of wetting of the entire
length of such a pier and the subsequent heaving of the pier.

Figure 4-17 Typical example of short pier foundation (insufficient pier
            length). Pier drilled into highly weathered claystone.
            Note: Adjacent Sonotube placed for underpinning.

It is important that the engineer also specify the minimum total pier
length in the foundation system to ensure that the piers are anchored
sufficiently deep in the unwetted zone of the bedrock.

Uniform pier diameter

After the pier hole is drilled, and during the placing of concrete, excess concrete is usually not removed from the top of the pier, resulting in a mushroom occurring at the top of the pier as indicated on Figures 4-18 and 4-19.

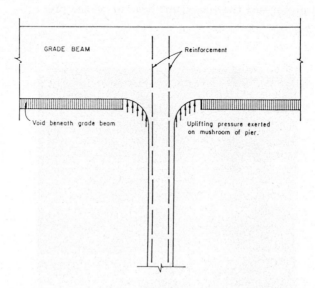

Figure 4-18   Effect of mushroom on the uplifting of the pier.

At times, the mushroom has been known to have a diameter three times the diameter of the pier.   Soil beneath the grade beams exerts direct uplifting pressure on the underside of the mushroom.   For a 12-inch-diameter pier with a 36-inch-diameter mushroom, the total area subjected to uplifting is 6.25 square feet.   For a moderately expansive soil having a swelling pressure of 10,000 psf, the total uplifting pressure exerted on the mushroom is about 62.5 kips.   This pressure alone will be sufficient to lift a pier provided it is not adequately anchored.

A great deal of emphasis has been placed on the formation of mushrooms on top of the piers.   To understand the function of the grade beam and pier system, it is be necessary to understand the commonly used construction method in this area.   Several construction methods have been employed in this area.   The one that renders the best results is shown in Figure 4-20.

The above method is to use a sonotube to extend the pier about four inches above grade, then pour the grade beam or the foundation wall on top of the piers. Verticel material is placed beneath the grade beam in direct contact with the Sonotube at both ends. In this manner, the formation of mushrooms on top of the piers can be avoided and a complete void beneath the grade beam can be established. The more common construction detail is to pour concrete in the pier holes flush with the ground surface. Place verticel as close to the pier top as possible, then pour the grade beams. Without the use of sonotubes, mushrooms invariably are formed. The

Figure 4-19 Typical mushroom 26 inches in diameter on top of a 12-inch
             diameter pier.

extent of the mushrooms depends on the stiffness of the upper soils, the disturbance of the surface soil during site grading and the pier drilling technique.

Theoretically, the presence of mushrooms has achieved two undesirable results. In the first place, the soil can exert direct swelling pressure on the base of the mushroom. In the second place, the dead-load pressure

applied on the pier is distributed over a larger area. The actual pressure
exerted on the end of the pier shaft is less than the designed value.

The above two unfavorable results are the main reasons why structural
engineers insist that the pier diameter must be true from top to bottom.
Many cases have been brought to the court pointing to the mushroom as the
primary cause of foundation movement.

In fact, it is obvious for those who are familiar with drilled pier
operation that the soil around the upper region of the pier hole is
disturbed by the slight eccentricity of the auger.   The amount of
disturbance depends on the type of soil. For sandy clays and clayey sands

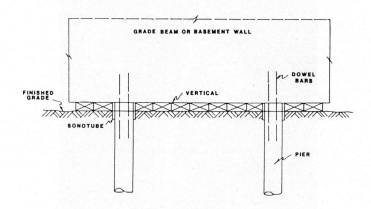

Figure 4-20   Void beneath grade beam and adjacent to pier.

(CL-SC), the amount of disturbance can be considerable. As a result, larger
mushrooms can be formed.   It is not unusual to find mushrooms with a
diameter twice as large as the shaft. Since the sandy soil possesses only
low swell potential, the presence of mushrooms actually does not contribute
to a great deal of pier uplift.

At the same time, if the soil consists of stiff clays with swelling
pressures on the order of 10 ksf as mentioned above, it is unlikely that the
drilling disturbance can affect the shaft diameter. The formation of large
mushrooms is, therefore, in this case unlikely.

The author believes that the importance of a true pier shaft has been
exaggerated. In most cases, the portion of pier uplifting force that can be

assigned to the mushroom effect is actually quite small. This does not mean that one should neglect the effort to obtain a uniform pier diameter. The use of sonotubes for the upper portion of the pier can be accomplished with little added cost and is highly recommended.

Pier reinforcement

Since the lower portion of the pier is anchored into bedrock by skin friction with the uplifting pressure acting upon the upper wetted portion of the pier, tensile stress develops within the pier. The maximum amount of tensile force developed can roughly be calculated as follows:

$$T = 2\pi rsd - P$$

Where T = total tensile force in lbs., and

P = total dead load pressure exerted on the pier in lbs.

For a 12-inch-diameter pier with a 15-kip dead-load pressure having four feet of unwetted length, and assuming that the skin friction of claystone is 2,000 psf, the total possible uplifting force is 10.1 kips or 89 psi. This stress can be taken by lightly reinforced piers; however, without reinforcement the pier will fail in tension. The location of the tension cracks is usually at the boundary of the wetted and unwetted portion of the pier. This zone is generally located at least three feet below grade beam; therefore, the normal dowel bars used in the piers do not provide the required resistance to tension. Generally, 0.6 to 1.0 percent reinforcement is sufficient, but there are cases when it is necessary to use as much as seven percent. Reinforcement of the full length of the pier is essential to avoid tensile failure. A typical tension crack in a pier is shown on Figure 4-21.

Air space

To prevent the lower soils from exerting uplifting pressure on the grade beams, it is essential that there be no contact between the soil and the grade beams. The required void space can be formed by the use of sand, cardboard, or other similar material which can be removed after the grade beam is poured. The most convenient method is by the use of a void-forming cardboard form known as "verticel." The cardboard material is wrapped in plastic and has adequate strength to support the concrete but will deteriorate after the plastic is punctured as shown on Figure 4-22. It is not necessary to remove the cardboard material after the completion of the grade beams. The cardboard form material also protects the backfill soils from plugging up the void space. Figure 4-23 indicates the location of the cardboard material.

The common practice is to ensure that the void space beneath the grade beam is well maintained so that soils do not have direct contact with the grade beam. The following were emphasized:

1. There should be sufficient void space so that when the soil swells, the space will not be filled.
2. The void material, such as verticel or J-void, should be strong enough to bear the weight of concrete and should be able to disintegrate under moist conditions.
3. The void space should come in contact with the pier so that there will be no area where the soil will have direct contact with the grade beam.
4. During the pouring of concrete, care should be taken so that concrete does not spill over the void material to form curtains which could possibly bear pressure.

Theoretically, all the above precautions are important for an effective pier and grade beam system. However, careful evaluation reveals that void space construction actually is not as important as most structural engineers have assumed.

The thickness of the void commonly used is four inches, which is the standard size for the verticel manufacturer. It is the author's opinion that a four-inch void should be sufficient to accommodate any type of soil. For example, a soil with a swelling pressure of 20,000 psf will expand 6% of its volume upon wetting. It requires a wetted depth of 5.5 feet to completely fill the void space beneath the grade beam. By this time, the swelling pressure exerted on the grade beam is zero, even if the soil comes in contact with the concrete.

Further expansion of the soil results in compressing the low density upper soil instead of exerting pressure on the grade beam. Swelling pressure is a direct function of dry density. As dry density decreases, the swelling pressure decreases proportionately. Again, in actual case, a four-inch void should be adequate under any conditions, contrary to many structural engineers' assumptions that when the void is filled a full swelling pressure is exerted on the grade beam.

Figure 4-21 Tension crack developed at approximately three feet below grade
             beams.

Figure 4-22 Deterioration of Verticel (void-forming cardboard) beneath
the grade beam.

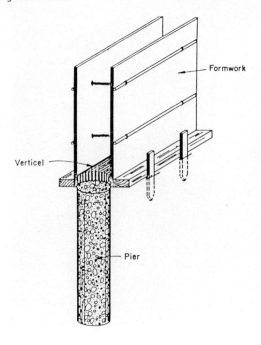

Figure 4-23 The use of Verticel (void-forming cardboard) beneath the grade
beam.

What about incomplete void under the grade beam? Let us assume that
the piers have a clear span of fifteen feet. Only 90% of the void is
complete. The remainder of the 10% area has soils coming in direct contact
with the grade beam. Assuming again that the soil has a swelling pressure
of 20,000 psf, the total uplifting force is

$$20 \times 15 \times 0.1 \times \frac{9}{12} = 22.5 \text{ kips}$$

Bear in mind that during site grading the upper soil has been loosened, and
the swelling pressure has, at most, half of the theoretical maximum, or on
the order of ten kips. This pressure, compared with the total uplifting
pressure exerted along the perimeter of the pier is relatively minor.

Again, the author is not advocating that a clean void space beneath the
grade beam is not important but thinks this factor has been overemphasized
as a primary cause of building distress.

Pier settlement

If piers are drilled deep into bedrock to provide the necessary
anchorage in a swelling soil area, pier settlement should not pose a
problem. Under normal pier load, the magnitude of pier settlement in
competent bedrock should be between 1/4 and 1/2 inches.

However, there are cases in which excessive settlement of the pier has
taken place resulting in severe cracking of the building. These cases
commonly occur in small projects where there is no construction control and
the pier driller determines the length of the pier and the amount of
penetration. If the pier is not bottomed on bedrock but instead on the
upper stiff clays, excessive settlement can occur. Stiff clays sometimes
have a strong resemblance to claystone bedrock, and a mistaken
identification of the bearing soil occasionally occurs.

A small amount of dry or plastic cuttings on the bottom of a pier hole
does not affect the bearing capacity of the pier. If, however, there are
one or two inches of soft soils at the bottom of the pier hole and concrete
is poured on the soft mud, then excessive settlement can occur. Such
phenomena usually take place in small-diameter piers (10 to 12 inches in
diameter) where the driller is unable to remove the mud accumulated at the
bottom of the pier hole by spinning the auger. Settlement of a single pier
can usually be bridged by the adjacent piers and is usually unnoticed.
Bearing capacity of the pier is usually much larger than the design-bearing

value. An exception is the chimney of a house. Chimney foundation pads are
generally supported by two to three piers. Settlement of one pier can cause
visible lifting and separation of the chimney as shown on Figure 4-24.

Void in pier shaft

Voids in pier shafts have been of great concern to foundation engineers
since the problem developed in a major structure in Chicago, Ill. Voids or
discontinuities in the pier shaft often result when a concrete which is too
stiff is used. In addition, voids are caused by the collapse of the casing,
by the squeezing of the soft formation, or by the hang-up of concrete in the
casing while being pulled.

Fortunately, piers drilled in expansive soil areas often encounter
stiff clays and the problem of squeezing of the soft formation seldom takes
place. However, cases are known where the reinforcing cage was inserted
into cased small-diameter piers with stiff concrete resulting in voids.

Small diameter piers on the order of 12 to 24 inches are usually
drilled without the use of casing. This is especially true in the case of
piers drilled into expansive, stiff clays. Consequently, the development of
discontinuities in shafts above bearing strata seldom takes place. However,
in numerous cases, voids have developed in the pier shaft and sometimes
complete discontinuity has taken place. The cause of such a defect is
generally due to inexperienced designers. Structural engineers design piers
using the criteria of long columns. Excessive steel and sometimes the use
of cages have prevented the flow of concrete. In other cases,
specifications called for the slump in concrete to be the same as that of
beams and columns. Stiff concrete with a slump less than four cannot expect
to flow into a reinforced 12-inch diameter pier even with the aid of
vibration.

Rise of ground water

The theory behind the drilled pier system is that there must be
sufficient dead-load pressure and the pier must be long enough so that the
lower part of the pier is embedded in a zone unaffected by moisture
change. Assuming that surface water will not penetrate more than about 15
feet, a 20-foot-long, heavily loaded pier should theoretically be free from
any possible movement.

Figure 4-24   Separation between chimney and house caused by pier
              settlement.

The exception is the rise of ground water where the once dry portion of
soil surrounding the pier becomes completely saturated. The skin friction
used for pier withholding is completely lost, and the pier lifts.

With the perched water condition, many of the defects and problems
developed in the drilled pier system can be explained.

In the first place, the interface between the pier and the soil could
have more affinity to water than the remainder of the soil mass. It is
almost certain that the pier punctures the seams and allows water to enter
the space surrounding the pier. As soon as this takes place, the skin
friction is lost and swelling along the pier shaft takes over.

It is obvious from the above discussion of perched water that the
drilled pier foundation system is no longer safe should perched water
develop in the area. The commonly accepted concept that the longer the
penetration of pier into bedrock, the safer will be the system, in this

case, may not be true. In fact, it may even have an adverse effect. The
use of a drilled pier foundation in areas where perched water is present
requires careful evaluation.

STATE-OF-ART DESIGN

   The mechanics of drilled pier foundation have never been fully
understood. The author realizes the defects and shortcomings of the present
system, yet a rational design has not been presented. It is admitted that
the present state-of-art knowledge on expansive soils has not provided us
with the solution. It is the author's opinion that we must return to the
basics. One of the basics is that the stability of the pier system is
controlled by the dead-load pressure. Dead-load pressure is the single
reliable element that resists uplift and is not affected by moisture
change. The key issue is dead-load pressure and a structural floor.

Structural floor

   For years, geotechnical engineers have been trying to save money for
the owner by advocating the use of a floating slab. The theory is that a
floating slab moves independently from the structural members so that floor
heaving and cracking do not affect the stability of the structure. Such
concepts have been accepted by most owners, developers and structural
engineers. Indeed, cost-wise, a substantial saving can be achieved by the
use of slab-on-ground. It is estimated that an additional cost of $3,000
must be added to a residential house should a structural slab be used. The
percentage of cost increase for a school building, office or warehouse can
be very much more.
   Indeed, the builder realizes the risk of slab-on-ground construction;
yet, the temptation of saving a considerable amount of construction costs
and in some cases beating the competition that is confronting him make him
decide to take the risk.
   Few realize that there is no such thing as a truly floating slab. All
slabs-on-ground adhere to the grade beams and when the slab heaves, it
exerts pressure to the grade beam affecting the structural stability. The
amount of pressure transmitted to the grade beam from a heaving slab can be
much greater than imperfect void space or mushroom of the piers. Slab
cracking parallel to the grade beam at a distance of about two feet from the
wall is strong evidence of stress transfer.

Further, no matter how carefully the carpenter forms the slab-bearing partition wall with a void space beneath them, there is some pressure transmitted to the upper structure through door jambs, sidings, staircase stringers, and others.

Improperly constructed subdrain systems tend to trap water around the perimeter of the basement. Such water may seep along the interface between the soil and pier and cause pier uplift. In short, the damage caused by slab heaving is not limited to floor cracking but also to structural damage.

The author believes it is time that the engineer stop emphasizing the economical aspect of a structural slab and point out to the builder the false economy of using slab-on-ground in highly expansive soil areas. The cost of repairing a badly damaged house and the adverse publicity derived from such events can easily be compensated by the use of a structural slab.

The most important advantage of using a structural slab is the increase of dead-load pressure on the piers. As explained under "Dead-Load Pressure," by the use of a structural floor slab, the dead-load pressure on the pier can be doubled. Thus, a relatively safe drilled pier foundation system can be accomplished.

It should be pointed out that a structural floor system must include an adequate space beneath the floor. A crawl space construction is most desirable. Otherwise, a space of at least 18 inches is recommended. The ground surface must be covered with plastic membranes to prevent dampness. The space should be well ventilated with artificial means to prevent condensation. A subdrain system in the crawl space may be necessary.

In Israel and in South Africa where highly expansive soils are encountered, engineers will not consider anything but a structural floor.

Dead-load pressure

It is difficult to exert sufficient dead-load pressure on piers supporting lightly loaded structures such as residential houses. For residential houses, the maximum dead-load pressure that can be exerted on piers with spacing of 15 feet is on the order of 30,000 psf. For a 12-inch diameter pier, this amounts to about 37,500 lbs. By using a structural slab system, such as precast concrete beams, the dead load per linear foot can be increased from 2,000 lbs. to 3,500 lbs. or about 68,000 psf. Obviously, in many areas of the house, only about half of the above calculated pressure can be exerted on the pier.

The author believes that if 40 to 70 ksf unit pressure can be exerted on the piers, many of the uplifting problems can be resolved. With years of experience on cracked houses, the author believes that the "magic number" is one-third. If the dead-load pressure exerted on a single pier exceeds one-third of the maximum possible uplift, the system is safe.

Pier length

As stated in the previous discussion, the traditional concept of increasing pier length to provide better anchorage into bedrock may not be applicable in the case of a perched water condition.

The presence of perched water in claystone bedrock is erratic, with no established pattern. However, it is believed that at lower depth, claystone bedrock has tightly compressed seams; the system is not continuous and not connected to exterior sources. If the piers can bottom on such stratum, then the anchorage effect can be achieved and is not affected by perched water.

The problem is to determine the elevation of such tight stratum, hence, to determine whether it is feasible to drill piers to reach the desirable layer. Careful core drilling should be able to disclose the condition of the claystone stratum at various depths. For residential construction, the cost of such investigation may not be economically feasible but for major structures such determination may be necessary.

For a successful drilled pier foundation in an area where a perched water condition may develop, the following should be observed:
1. The pier should have a high minimum dead-load pressure, 40,000 to 70,000 psf.
2. The pier should penetrate into bedrock. The amount of penetration depends on the characteristics of bedrock.
3. A structural floor slab with adequate space beneath the slab is necessary.
4. Usual precautions such as adequate void space beneath the grade beams, the proper formation of the top of the piers and the proper ventilation of the space beneath the floor should be observed.
5. A subdrain system with proper slope and proper outlet is required.
6. Good surface drainage as prescribed by engineers should be maintained by the owners.

With the above system, the problems that are experienced in building on expansive soils can be greatly reduced. The recommendations given can by no means be treated as new approaches. Engineers have understood such practices for a long time; still, most of them do not want to deviate from the traditional practice.

We must admit that a solution to the drilled pier system is not available. We have little knowledge on the mechanics of the performance of drilled piers in expansive soils. Large-scale field experiments on drilled piers are almost non-existent. In the academic arena, little attention has been directed to the behavior of partially saturated clays, the magnitude of uplift pressure caused by soil expansion on piers, the magnitude of skin friction and the value of skin friction when wetted. A great deal of research both theoretically and experimentally will be necessary before a rational answer can be expected.

REFERENCES

[4-1]   Woodward, R., Gardner, W.S. and Greer, D.M., "Drilled Pier Foundation," McGraw-Hill Book Company, 1972.

[4-2]   Jubenville, David M. and Hepworth, Richard, "Drilled Pier Foundation in Shale, Denver, Colorado Area," Proceedings of the Session on Drilled Piers and Caissons, ASCE/St. Louis, MO, 1981.

[4-3]   Parcher, J.V. and Liu, P.C., "Some Swelling Characteristics of Compacted Clays," Journal of the Soil Mechanics and Foundation Division, ASCE, Vol. 91, pp. 1-17.

[4-4]   Mohan, D. and Chandra, S., "Frictional Resistance of Bored Piles in Expansive Clays," Geotechnique, Vol. XI, No. 4, pp. 294-301.

[4-5]   Seed, H.B., Mitchell, J.K. and Chan, C.K., "Studies of Swell and Swelling Pressure Characteristics of Compacted Clays," Highway Research Board Bulletin 313.

[4-6]   O'Neill, Michael W. and Poormoayed, Nader, "Methodology for Foundations on Expansive Clays," Journal of the Geotechnical Engineering Division, ASCE, Vol. 106, No. GT 12, Dec. 1980.

[4-7]   Army Engineer Waterways Experiment Section, "Engineering Properties of Clay Shale," Aug. 1974.

[4-8]   Teng, W.C., "Foundation Design," Prentice-Hall, Inc. 1962.

Chapter 5

FOOTING FOUNDATIONS

INTRODUCTION

Footing foundations can be successfully placed on expansive soil provided one or more of the following criteria are met:
1. Sufficient dead-load pressure is exerted on the foundation.
2. The structure is rigid enough so that differential heaving will not cause cracking, or
3. The swelling potential of the foundation soils can be eliminated or reduced.
The types of footing foundations are shown in Figure 1-1.

CONTINUOUS FOOTINGS

The most common type of foundation for lightly loaded structures is continuous footings. Local building codes sometimes specify the minimum allowable width of footing as 20 inches, which is not applicable for footings which are to be placed on expansive soils. To concentrate sufficient dead load pressure on expansive soils, the width of the footing should be as narrow as possible.

It should be noted that continuous spread footings cannot be expected to function well in highly expansive soil areas. The use of this system should be limited to soils with a low degree of expansion: those having a swelling potential of less than one percent and a swelling pressure of less than 3,000 psf.

Generally, the dead-load pressure exerted on a continuous foundation is low and in the following ranges:

Single story schools	2,000 to 4,000 lb. per ft.
Basement house	1,000 to 1,500 lb. per ft.
Butler type building	< 500 lb. per ft.

To ensure that a dead-load pressure of at least 1,000 psf is exerted on the soil, it is necessary to use very narrow footings, in most cases less than 12 inches wide.

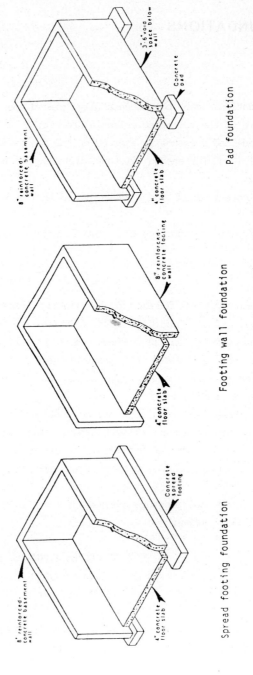

Figure 5-1 Types of spread footing foundation

It is obvious that with soils of high swell potential, continuous footings cannot be successfully adopted. However, an alternate foundation system may prove to be economically unjustified. A statistical survey of 180 houses was conducted in a suburb of Winnipeg, Canada [5-1]. These houses were supported on shallow foundations resting on highly expansive clays. The unit-bearing pressures on the subsoils beneath the perimeter walls were on the order of 1,000 psf. The survey revealed that 85% of the houses had developed cracks in walls and ceilings. Only five percent of the damaged houses was classified as major. The average cost of house repairs was estimated at about $60 per year. The survey concluded that from an economic point of view replacing shallow continuous footings with pile foundations or other systems was not justified.

In China where an alternate foundation system is not possible, engineers recommend placing the footings at lower depths [5-2]. Shallow foundations suffer more damage than deep foundations. It has been generally accepted in China that surface moisture will penetrate the subsoils for a depth of five feet. In arid regions, this depth is increased to eight feet.

In accordance with the Chinese "Building Code in Expansive Soils Regions," the depth of foundation in relation to severeness of expansion is as follows:

Degree of Expansion	Depth of Foundation
Class I - very severe	6.5 ft.
Class II - severe	6.5 ft.
Class III - moderate	5.0 ft.
Class IV - light	3.3 ft.
Class V - very light	1.8 ft.

The above is based on distress experienced on brick and masonry structures. For instance, Class V expansive soil will not cause cracking of brick and masonry structures. At the same time, regarding similar structures placed on Class I expansive soil, the differential movement can exceed 2.5 inches with cracking in excess of three inches wide. The suggested depth of foundation may limit the differential movement to 0.4 inches.

The author is of the opinion that under the following conditions, the use of continuous footings may be considered.

1.   When the bearing soils are not highly expansive.   (Typically, illite
     instead of montmorillonite).
2.   When the foundation beams are heavily reinforced.
3.   When the development of a perched water condition is remote.
4.   When pier drilling facilities are not available.
5.   When the superstructure consists of wood-frame construction.

Wall footings

     Engineers often specify the erection of basement walls directly on the
soil without the use of footings. This reduces the bearing width to about

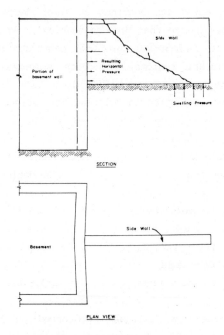

Figure 5-2. Plan and section of foundation wall bearing directly on
             expansive soil. Note pressure distribution.

nine inches and increases considerably the unit dead-load pressure exerted
on the soils. Such a concept is sound from the expansive soil standpoint.

Figure 5-3. Foundation wall bearing directly on expansive soil without
             footings. Heaving has pushed the side wall toward the basement
             wall resulting in heavy bowing of the basement wall.

However, care should be exercised to ensure the rigidity of the system by
checking the following conditions before construction begins:

1. Determine if there are any soft pockets in the excavation that may
    introduce settlement,

2. Ensure that there is sufficient continuous reinforcement in the
    foundation wall to provide rigidity, and

3. Make sure that the walls are properly restrained against earth
    pressure.

An extreme case recently occurred which involved a wall bearing
directly upon expansive soil. The upper wall heaved and imparted horizontal
pressure to the basement wall resulting in heavy bowing of the wall even
before the house was completed (Figures 5-2 and 5-3).

Box construction

The use of heavy reinforcement in the foundation wall can protect the
structure from cracking due to differential heaving. The average height of
a concrete basement wall is six feet. Such walls can span an unsupported
length of at least ten feet and can therefore tolerate considerable
differential movement without exhibiting cracks. Weak points do appear at

points of discontinuity, such as doors, deep windows, and change of elevation.

Box construction is based upon the principle that there is no discontinuity of structure; therefore, there are no weak sections. Box construction is economical for structures having simple configurations. For split-level residential houses or basements with walk-out doors, such construction is more difficult. Consideration should then be given to the use of a construction joint to separate the structure into two or more units. Each unit will then act independently, and differential movement can be confined to the joints.

S. Shraga and D. Amir [5-4] reported the use of box construction in Kibutz Gat, Israel, where the structure consists of two reinforced concrete boxes each about 22 by 35 feet in dimension. The structure did not exhibit damage after 17 years despite the considerable differential movements of up to five inches between the corners of individual boxes. Shraga and Amir concluded that box construction can structurally withstand movement and tension without cracking.

Masonry bricks and cinder blocks cannot withstand movement and should not be used for foundation walls. The small saving derived from using masonry construction instead of concrete foundation walls may later result in heavy loss of property in the event of foundation movement.

Reinforced brickwork has been widely used in South Africa. D.L. Webb [5-5] reported the use of reinforcement in the external wall panels between joints to resist bending stresses and shear stresses resulting from foundation movement. The arrangement is shown on Figure 5-4.

PAD FOUNDATIONS

The pad foundation system consists essentially of a series of individual footing pads placed on the upper soils and spanned by grade beams. The principle of a pad foundation system is similar to that of a drilled pier foundation in that the load of the structure is concentrated at several points, the difference being that the pads bear on the upper soils and skin friction is not involved.

Under the following conditions, the use of a pad foundation system can be advantageous:

1. Where bedrock or bearing stratum is deep and cannot be economically reached by drilled piers,

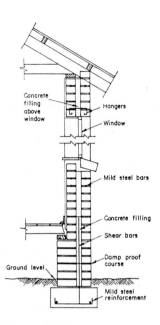

Figure 5-4. Section through externally reinforced brick wall.
            (After D.L. Webb).

2. Where the water table or a soft layer exists preventing the use of a
    friction pier,
3. Where the upper soils possess moderate swell potential, and
4. Where the bearing capacity of the upper soils is relatively high.

Design

By loading an expansive soil so that the pressure exerted on the soil
is greater than the swelling pressure of the soil, heaving movement can be
prevented.

By using an individual pad foundation system, it is theoretically
possible to exert any desirable dead-load pressure. Actually, the capacity
of the pad is limited by the allowable bearing capacity of the foundation
soils. If a pad is founded directly on bedrock, maximum allowable soil

pressure will not pose a problem. However, if the pads are placed on stiff swelling clays, the maximum bearing capacity of the pad is limited by the unconfined compressive strength of the clay. Generally, the maximum bearing capacity should be about 5,000 psf. Consequently, the practical dead-load pressure that can be applied to the pad is about 3,000 psf (considering the ratio of dead and live load to be about two to three). Occasionally, pads founded on clay are designed to withstand a dead-load pressure as high as 5,000 psf. With this limitation, an individual pad foundation system can only be used in those areas where the soils possess only a medium degree of expansion with volume change—on the order of one to five percent and a swelling pressure in the range of 3,000 to 5,000 psf.

To allow for the concentration of dead-load pressure on the individual pads, a void space is required beneath the grade beams and should be constructed in the same manner as the grade beams and the pier system (Figure 5-5). Figure 5-6 shows a grade beam and pad foundation system that failed because the dead-load pressure was not sufficient to prevent the heaving of the foundation soils.

Peck, Hanson & Thornburn stated in the second edition of Foundation Engineering that, "Swelling can be prevented only in a localized zone beneath the footings or piers where the stress induced by the foundation are concentrated." This is shown on Figure 5-7.

At a comparatively shallow depth beneath the foundation, the intensity of added stress is small and swelling may occur below this level, even if it is entirely prevented above. In the area between the footings, swelling is undiminished.

Deep pads

In areas where the layer of swelling soils is relatively thin, deep individual pads placed on nonswelling soil can be economically used. A typical example is the situation in which two to three feet of swelling clays are underlain by sand and gravel or by nonswelling bedrock such as granite or sandstone.

Pads placed as deep as five feet below the ground surface can be economically used in areas where drill rigs are unavailable. Care should be exercised to ensure that uplifting pressure will not be exerted on the sides of the pad. The excavation should be larger than the footing pad; the space between the concrete and the soil should be filled with loose backfill.

Figure 5-5. Grade beams and pads constructed with void space between pads.

The use of a deep pad system usually applies to construction areas where the problem soil ranges in thickness from zero to five feet. In such cases, it is desirable to place all footing pads on uniform nonswelling soils.

In those parts of the world where hand labor is inexpensive and drilling equipment not readily available, the use of a deep pad system can be an advantage from a cost consideration.

Interrupted footing

Interrupted footings are used in conjunction with a wall footing system. With foundation walls bearing directly on swelling soil, the maximum unit dead-load pressure exerted on the soil is about 2,000 lbs. per ft. By placing a void space at intervals, the bearing area will be decreased, thus increasing the dead-load pressure. In this manner, the dead-load pressure exerted on the soil can be easily doubled.

This principle of interrupted footings has been successfully applied to the correction of cracked buildings having a continuous footing

Figure 5-6. A typical crack which developed in the basement of house
             founded with grade beams and individual pads. The dead-load
             pressure was not sufficient to prevent pad uplift.

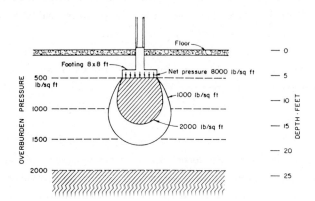

Figure 5-7. Diagram illustrating influence on swelling of high contact
             pressure beneath footing. If net pressure at base of footing
             is 8,000 lb/sq ft and swelling pressure at zero volume change
             is 2,000 lb/sq ft, swelling will be prevented within shaded
             areas only. (After Peck, Hanson & Thornburn)

foundation. By introducing some void space beneath the footings, the dead-load pressure can be substantially increased, thus preventing further foundation movements. (See Case Study IV for illustrations)

FOOTINGS ON SELECTED FILL

The removal of natural expansive soils and their replacement with nonexpansive soil is the most obvious method of preventing structural damage due to soil heaving. In a few cases, it may be possible to completely remove the expansive strata, thus eliminating the heaving problem. In most cases, the expansive material extends to a depth too great to allow complete removal and backfill. The problem is then to determine the amount of excavation and the type of backfill required to prevent heaving. Detailed discussion is given in Chapter 8 under "Soil Replacement."

Probably the most important single factor affecting the success of footings placed on selected fill is the drainage control used during construction. If the excavation is wetted excessively before the placement of the selected fill, the imbibed moisture in the soil causes the soil to swell and heave and exert pressure against the selected fill, resulting in severe damage to the structure. Many school buildings have been successfully placed on selected fill, both for the entire system and for slabs alone. Coincidentally, there have also been failures when such construction has been used, mainly because the site was flooded during construction.

For success in placing footings and slabs on selected fill, the following precautions should be observed:

1. There should be at least three feet of selected fill beneath the bottom of the footings and slabs.
2. The fill should extend beyond the building line for a distance of at least ten feet in every direction.
3. The fill should consist of nonexpansive soil, preferably impervious and granular.
4. The fill should be compacted to at least 90 percent standard Proctor density for supporting slabs and 100 percent standard Proctor density for supporting footings.
5. Before the placement of fill, care should be taken to avoid the excessive wetting of natural soils.

MAT FOUNDATION

Mat foundations, sometimes referred to as structural slab-on-ground or reinforced and stiffened slabs, are considered to be both a load supporting as well as a separating element. The slab receives and transmits all of the structural load to the underslab soils. The slab should be designed to resist both the positive and the negative moments. Positive moment includes that induced by both dead and live-load pressure exerted on the slab. Negative moment consists mainly of those pressures caused by the swelling of the underslab soils. Since the swelling pressure in an expansive soil area can reach many thousand pounds per square foot, negative moment consideration generally controls the design of the mat foundation. If all structural elements are to be placed on a stiffened slab, then slab movement will not affect the stability of the structure. There could be tilting of the mat, but the performance of the building would not be structurally affected. Such a concept has been studied by the "Building Research Advisory Board" [5-6]. A study on the work in the rocky mountain areas indicates that there are limitations to the use of such a system as follows:

1. The success of such a system so far is limited to moderate swelling soil areas.

2. The configuration of building must be relatively simple.

3. The load exerted on the foundation must be light. Past performance has been limited to residential construction.

4. Single level construction is required. It would be difficult to apply such construction to basement houses with an attached garage or split level houses.

BRAB design

The Building Research advisory board design concept of a mat foundation is generally based on the following parameters:

1. Soil conditions,

2. The support index, and

3. The dead and live load acting on the slab.

From the above parameters, the designer must develop a structure capable of satisfying the shear, bending moment, and deflection conditions.

The first step in the design is to determine the support index. The support index is based upon the climatic rating, the plasticity index, and the length-to-width ratio of the foundation.

It is assumed that the moisture content in the soil is affected by climatic conditions, and the volume change of the soil is affected by the moisture content. Likewise, both the swelling and volume change of the soil are affected by the moisture content. Likewise, both the swelling and the settlement of the soil are affected by climate. A study of weather data indicates that the yearly annual precipitation, distribution of precipitation, frequency of precipitation, duration of precipitation, and amount of each precipitation all affect the consistency of climate. Based on data obtained from 122 weather stations, the U.S. National Weather Service has developed information which has been transformed into frequency isolines on a map of the continental United States as shown on Figure 5-8. From this figure, the climate rating C_w is selected. For instance, in Colorado, the climate rating is between 20 and 25.

The second major factor necessary for design is the support index which is directly related to the climate factor and soil properties. Figure 5-9 shows the relationship of the various properties. The soil properties are related to the Atterberg limits, the percent swell in the PVC meter, and the swell index. Of these, the swell index is the most reliable factor of the three for predicting the potential volume change of the foundation soils.

Soils with identical plasticity indices exhibit greatly varying swell potentials. Also, the PVC meter is based on testing soils in a remolded state which can materially differ from that in the undisturbed state.

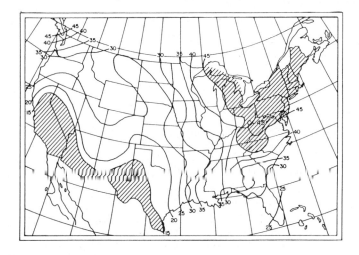

Figure 5-8. Climatic ratings C_w for Continental United States.
(After Federal Housing Administration).

To obtain the swell index, the percentage swell for a specific soil stratum should be obtained through swell tests using conventional consolidometer test equipment on undisturbed soil samples and pressure corresponding to the in situ overburden pressure plus the average of the total dead and live loads on the slab. The undisturbed samples should be obtained under soil moisture conditions representative of conditions prevailing at the time of construction.

With the swell index or the percent of swell under specific loading conditions and the climate rating determined, the support index can then be determined from Figure 5-9. The design of the stiffened slab section will be based on the value of the support index.

The Building Research Advisory Board procedure assumes a relationship between a weighted plasticity index and a climate rating to determine the degree of support the soil is expected to provide to the slab. The author does not believe the existence of the PI and C_w relationship. In fact, it is doubtful that climate ratings should enter into the design at all. This means that the only soil information used in the design is the plasticity index. Since it is well established that the plasticity index alone is not sufficient to control the characteristics of swelling soil, it is questionable that such sparse subsoil information has been fed into the design.

In a review of stiffened slab-on-ground construction, Warren Wray [5-7] pointed out that the differential swell is the controlling factor governing the success of such construction. If the soil beneath the slab were to swell uniformly, no distortion would be caused in the slab. Distortion occurs when the supporting soil swells non-uniformly or differentially.

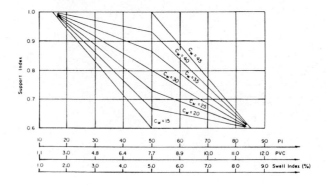

Figure 5-9.   Support index C based upon criterion for soil sensitivity and climatic rating C_w. (After Federal Housing Administration).

Texas Design

R.L. Lytton and J.A. Woodburn [5-8] of Texas A & M University have performed considerable research on the design procedure for stiffened mats on expansive clay. Lytton and Woodburn determined the support index algebraically from the average foundation pressure, subgrade modulus, maximum expected differential heave of the soil and the mounded area. Lytton and Woodburn prepared a nomograph for determining the support index as shown on Figure 5-10.

With the support index determined, the design of the mat foundation is within the realm of a structural engineer. A typical mat foundation design is shown on Figure 5-11.

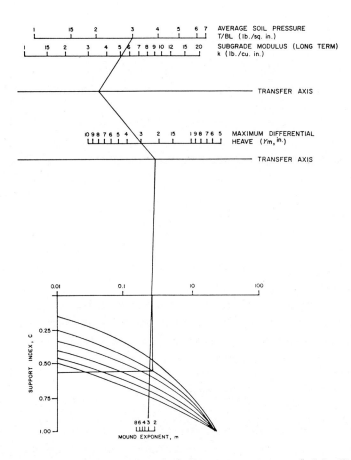

Figure 5-10. Support index nomograph (After R.L. Lytton and J.A. Woodburn).

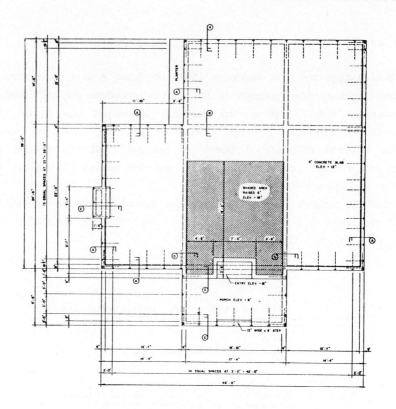

Figure 5-11. Typical mat foundation design (Sheet 1 of 2).

PTI Design

        Design procedures recommended by the Post-Tensioning Institute [5-9]
have been widely used by engineers throughout the United States. The design
procedure can be summarized as follows:

(1) Select a Thornwaite Moisture Index from Figure 5-12.

(2) Obtain an estimate of edge moisture variation distance e_m, for both
     edge lift and center lift loading condition from Figure 5-13.

(3) Calculate the activity ratio of the soil

$$A_c = \frac{PI}{(\% \text{ passing \#200 sieve})}$$

Where A_c = Activity ratio
       PI = Plasticity index

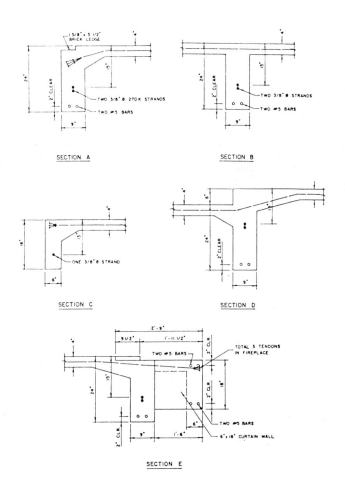

Figure 5-11. Typical mat foundation design. (Sheet 2 of 2).

(4) Calculate the cation exchange activity to determine the soil mineral
type.   The clay in most cases may be conservatively classified as
montmorillonite.

(5) Determine the depth to constant soil suction.   Constant soil suction
can be estimated from Figure 5-14.   For most practical applications,
the design soil suction value will seldom exceed a magnitude of
pF = 3.6.

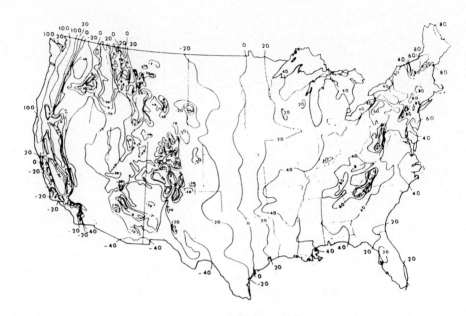

Figure 5-12. Thornthwaite moisture index distribution in the United States

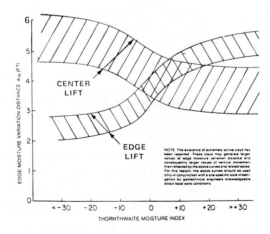

Figure 5-13. Approximate relationship between Thornthwaite Index and
moisture variation distance.

(6) Moisture velocity can be approximated by using a velocity equal to one-half of the Thornwaite Moisture Index for the construction site, converted to inches/month. This value shall not be assumed to be less than 0.5 in./month and the maximum moisture velocity shall be 0.7 in./month.

(7) Using the values of the edge moisture distance variation, e_m, percent clay, predominant clay mineral, depth to constant suction, soil suction pF and velocity of moisture flow, it is possible to determine the corresponding soil differential movement, Ym. This is accomplished by a series of computer programs based on the permeability of clays and the total potential of soil water.

The above design procedure leads to the basic soil-structure interaction models as shown on Fig. 5-15. If the soil beneath the slab experiences a change in its moisture content after construction of the slab, it will distort into either a center lift mode or an edge lift mode. Structural design of the system follows by using the differential soil movement value Ym.

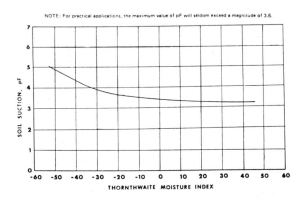

Figure 5-14. Variation of constant soil suction with Thornthwaite Index

The primary criteria which the structural engineer requires from the geotechnical engineer for the design of post-tensioned slabs based on PTI design procedures consist of the differential swell Ym for edge lift and center lift conditions. Ym values are estimates of differential movements one would expect due to the expansive soils on the site. The PTI procedure results in selecting differential swell values based on empirical correlations with climate and clay mineralogy to an accuracy of 0.001

inches, as all practicing engineers realize that the accuracy may be only within one inch.

The structural engineers, on the other hand, notice the impact variation in several tenths of an inch that the differential swell values have on structural design. They cannot accept a round-off value of, say, one to 1.5 inches.

The author questions whether the soil parameter used in the procedure can fit in with the precise structural parameter suggested in the procedure.

Behavior

Stiffened slab construction has been widely used in southern Texas where moderate swelling soils are encountered. The so-called waffle slabs

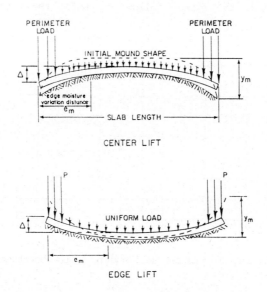

Figure 5-15. Soil-structure interaction models

have been in use in San Antonio for more than 25 years and are also required for F.H.A.-sponsored construction in Montgomery, Alabama.

In Denver, such a foundation system was first considered in 1970. Subsequently, 52 houses were built using waffled slabs in West Field Park,

Jefferson County. The soils are generally stiff clays with a plasticity index ranging from a low of 3.6 to a high of 32.1. The swell potential ranged from a low of one to a high of 5.5 percent under loads ranging from 500 to 1,000 psf. The soils are considered to have moderate swell potential.

In the same year, twelve houses were built using the stiffened slab system in Lake Arbor Subdivision in North Denver. The soils in the Lake Arbor area possess much higher swell potential than that of West Field Park, exhibiting swelling pressure as high as 10,000 psf. Typical design and construction details of these houses are given in Figures 5-16 through 5-21.

A survey of these houses was made in 1974 and their condition was excellent. None of these houses exhibited noticeable cracks. Differential elevation between the opposite corners of the house in some cases reached one inch, but distress, either in the interior or the exterior, had not taken place. In the same Lake Arbor area, some other houses have required replacement of their basement floor slab three times within four years. This indicates that the stiffened slab system used for the twelve houses appears to be highly successful.

Figure 5-16. Trenching the cross-beams.

Figure 5-17. Placing reinforcement.

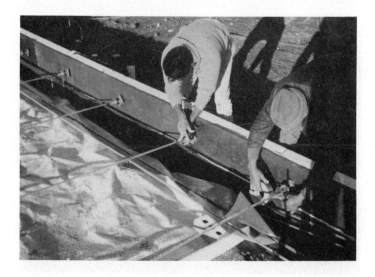

Figure 5-18. Post-tensioning.

Figure 5-19. Placing concrete.

Figure 5-20. Completed mat.

In recent investigations on townhouses in southwest Denver, the upper clays had a liquid limit ranging from 50–60% and a plasticity index from 30–40%. Claystone bedrock is at a shallow depth with a liquid limit ranging from 50–70% and a plasticity index of about 50%. The swell potential reached as high as 7% under a surcharge load of 1,000 psf with swelling pressure as high as 30,000 psf. Shortly after the completion of the complex, movement appeared in the pattern of wall cracking, door binding and ceiling distortion. Most of the distress appeared to be related to edge lifting. The exterior of the building and the typical movement are shown on Figures 5-22 and 5-23.

Similar activity has taken place in San Antonio where a large number of houses have suffered mild-to-very-severe damage. It was reported that more damage has taken place with the post-tensioned slabs. It is entirely possible that construction techniques have not been closely supervised.

Figure 5-21. Interior partitions.

Figure 5-22. Exterior of structure.

Figure 5-23. Movement of brick course above post-tensioned slab.

Evaluation

As stated in the opening remarks of the advisory board [5-6], "It is recognized that experience and the state of engineering knowledge are such that precise answers to many of the problems posed must, of necessity, be considered beyond attainment in the immediate foreseeable future. Nevertheless, the approach recommended herein is considered to be sufficiently valid to warrant application now."

General evaluation of the mat foundation system is as follows:

(1) Based on our experience and performance relating to the Denver project, the stiffened slab system of construction can be successfully applied to low-to-moderate swelling soil areas. Much research is required to determine and understand the many variables, especially the relationship of swelling characteristics with a support index. As discussed previously in "Design," the support index should be related to the swelling pressure. Thus, the loading condition can be eliminated from the design as well as the climatic rating.

(2) Since differential swell is the key to the design of this system, every effort should be made to minimize such movement. The author suggests that all mat foundations should be placed on top of at least three feet of compacted nonexpansive fill. Heaving of the expansive soils beneath the fill cannot be avoided, but differential swell with the slab can be greatly reduced.

(3) In the Denver area, the increased cost of using stiffened slabs rather than the conventional pier and grade beam system was about 50 cents per square foot. This amounted to an increase in cost of about $750 per house which is small when compared to the problems and rehabilitation costs encountered where the stiffened slab system had not been used.

(4) One must realize that with a mat foundation system, remedial repair is difficult, if not economically feasible. In the San Antonio area, many houses are now under or have gone through extensive repair. The author questions whether the remedial repair can be successful.

TUNNEL FOUNDATION

Very little literature has been devoted to problems in tunneling through expansive soils. Most deep tunnels were bored through rocks with no swelling problems. In boring tunnels through expansive shale, such as the experience of Pierre & Mancos, engineers must take swelling problems into

consideration.   The issues normally involved are the coefficient of earth
pressure, the coefficient of lateral pressure, effective overburden pressure
and elastic and non-elastic rebound.   With swelling soils, the issue is
further complicated by vertical swelling pressure under the natural state
and under the air-dried state, as well as horizontal swelling pressure.
These pressures can be acting independently and can be superimposed.

Mechanics of swelling

The assumed mechanics of stress alteration around the tunnel are as
follows:

(1) Immediately after the completion of the borehole, elastic rebound takes
    place.   The magnitude of the rebound depends on the duration of
    borehole exposure.   Non-elastic rebound can take place gradually.

(2) After the conduit has been pushed into the borehole and the space
    between the conduit and rock is filled with grout, active earth
    pressure is exerted on the conduit.   Water from the grout or other
    sources will cause shale to expand and swelling pressure will also act
    upon the conduit.

(3) The magnitude of at-rest earth pressure can easily be determined by
    assuming the angle of internal friction.   The swelling pressure will
    vary depending on the following:
    (a) The amount of available water.
    (b) The ratio between vertical swelling pressure and horizontal
        swelling pressure.
    (c) The plasticity and initial moisture content of the shale.

(4) The realistic pressure exerted on the conduit can be much smaller than
    the above assumed value.   This is due to the following:
    (a) The rebound increases the void ratio and, consequently, decreases
        the swelling pressure.
    (b) Irregular clearance between the conduit and borehole.
    (c) Flexibility of the conduit.

(5) It is estimated that real pressure exerted on the conduit can be only
    50% of the theoretical value.

(6) According to Mesri [5-10], the extent of the development of swelling
    pressure is about twice the tunnel diameter.   More research is required
    to evaluate this assumption.

(7) It is the author's contention that the most critical period of stress
    concentration on the tunnel is when lateral swelling pressure exerts

its full force. This period is shortly after construction. As the stress condition reaches equilibrium, the tunnel lining is stabilized.

Effective stress

Effective stress is the stress actually exerted on the tunnel lining. This value must be estimated so that lining thickness can be ascertained.

Mesri stated that if the volume increase of shale is assumed to be controlled by the effective octahedral stress, then the field and laboratory values of effective octahedral stress have to be equal in order to prevent swelling. Such a consideration leads to the following equation.

$$\sigma'_s = [\frac{1+ 2K_o}{1+2K}] \; \sigma'_{vo}$$

Where σ'_s = Effective stress

σ'_{vo} = Effective overburden pressure

K = Coefficient of lateral pressure

K_o = Coefficient of earth pressure

Mesri claimed that for highly overconsolidated swelling shale, K_o can be much higher than one.

The magnitude of lateral swelling pressure can reach three to ten times as high as the vertical swelling pressure under partially saturated conditions, as stated in Chapter 2. Lateral swelling pressure decreases gradually as the soil approaches complete saturation and finally reaches equilibrium of a value equal to that of the vertical swelling pressure.

The swelling pressure under the rebound condition simulates the condition of the tunnel after the completion of the boring. Tests indicate that the swelling pressure as measured form the rebound curve is only about 70% from that measured from the compression curve.

Construction precautions

Based on the above research, it appears that the following construction precautions should be exercised to minimize the effect of swelling soil on the tunnel conduit:

(1) Every effort should be made to avoid excessive wetting of the soils around the tunnel. Excessive use of water in the grouting operation should be avoided.

(2) It is advisable to leave the tunnel boring open for as long a period of time as permitted. This is to allow the elastic rebound to take place and at least part of the non-elastic rebound to occur.

(3) It may be necessary to install a drainage facility along the tunnel so that any water which has entered the tunnel can drain out.

REFERENCES

[5-1]   Domaschuk, L., etc., "Performance of House Foundations on Expansive
         Soils, 5th International Conference on Expansive Soils, Australia,
         1984.

[5-2]   Hou, Shitao, "Types of the Expansive Shrinkage Soil in China and
         Their Engineering Geological Characteristics," Hubei Comprehensive
         Survey and Prospect Institute, 1979.

[5-3]   Chen, Fu Hua, "The Distribution and Characteristics of Expansive
         Soils in China," 5th International Conference on Expansive Soils,
         Australia, 1984.

[5-4]   Shraga, S. Amir, D., and Kassiff, G., "Review of Foundation Practice
         for Kibbutz Dwelling in Expansive Clay." Proceedings of the Third
         International Conference on Expansive Soils, 1973.

[5-5]   Webb, D.L., "Foundations and Structural Treatment of Buildings on
         Expansive Clay in South Africa," Second International Research and
         Engineering Conference on Expansive Clay Soils, Texas A & M Press,
         1966.

[5-6]   "Criteria for Selection and Design of Residential Slabs-on-Ground,"
         Building Research Advisory Board.

[5-7]   Wray, Warren K., "Analysis of Stiffened Slab-on-Ground Over
         Expansive Soil," 4th International Conference on Expansive Soils,
         Denver, 1980.

[5-8]   Lytton, R.L. and Woodburn, J.A., "Design and Performance of Mat
         Foundations on Expansive Clay," Proceedings of the Third
         International Conference on Expansive Soils, 1973.

[5-9]   "Design and Construction of Post-Tensioned Slab-on-Grade," Post-
         Tension Institute, 1980.

[5-10]   Mesri, G., "Report on Laboratory Measurement of Swelling Pressure
         for Taylor Shale and Time-Rate of Swelling Pressure Development
         Against a Tunnel Lining," 1984.

Chapter 6

SLABS ON EXPANSIVE SOILS

INTRODUCTION

Slab-on-ground construction, when on expansive soils, is a very difficult structural entity to control. In the category of slabs are interior floor slabs, exterior sidewalks or aprons, and patio slabs. Generally, floor slabs do not support any appreciable live load, and the dead load actually exerted on the slab is small. Consequently, movement of the slab is to be expected when the underslab moisture content increases, and it should be designed accordingly. The movement of slabs not only presents unsightly cracks but, in most cases, also directly affects the stability of the structure.

SLAB-ON-GROUND

Concrete slabs, placed directly on the ground, are much less expensive than structural floor slabs or "crawl space" type construction. This is especially true where basement construction is involved. Since 1940, most of the residential houses, school buildings, industrial, and warehouse structures called for the use of slab-on-ground construction. It was not until the discovery of the expansive soil problem that engineers began to question the wisdom of using slab-on-ground construction.

Types of slab-on-ground

A slab-on-ground, sometimes referred to as a slab-on-grade, is a concrete slab placed directly on the ground with little consideration given to its structural requirements. These slabs are constructed both with or without reinforcement.

The unreinforced slabs are generally constructed in residential houses or where a light floor load is expected. The limits of the length of the unreinforced slab are based upon the amount of shrinkage cracking control desired. Normally, shrinkage cracks are controlled by designed-weakened plane joints.

A lightly reinforced slab is normally reinforced with a temperature control as a prime design factor. The Portland Cement Association [6-1] recommended the use of a 4-inch thick slab reinforced with 6x6 - 10/10 mesh or No. 3 bar at 24-inches on center each way for slabs placed in moderately swelling soil areas. For high swelling soil areas, the Association recommended the use of 6x6 - 6/6 mesh or No. 3 bar at 18 inches on center each way.

The choice between an unreinforced slab and a lightly reinforced slab depends upon the subsoil conditions as well as the loading conditions. Reinforcement in the slab will reduce the opening of temperature cracks but will not prevent cracking of the slab caused by heaving of the underslab soils.

In 1968, a report concerning residential slab-on-ground construction was prepared by the Building Research Advisory Board [6-2] for use by the Federal Housing Administration which provided criteria for the selection and design of residential floor slabs. The report recommended the use of unreinforced concrete slabs for firm, nonexpansive soils. Nominal reinforcement was recommended where the subgrade may undergo slight movement. Both reinforced and unreinforced slabs are considered to have the limiting function of separating the ground from the living space.

Slab-on-ground construction on expansive soil will always pose a cracking and heaving problem unless the subgrade soils are treated or replaced. In commercial buildings such as warehouses and storage areas, where floor loads as high as 3,000 psf are expected, special design is required, not only from the standpoint of expansive soils, but also to maintain the structural integrity of the building. Minor floor cracking of slab-on-ground construction is difficult if not impossible to prevent.

Slab movement

In expansive soil areas, floor movement is invariably associated with the increase of moisture content of the underslab soils. The source of water that enters into the underslab soils can generally be associated with the following:

1. Rise of ground water, usually perched water, can cause excessive swelling. Heaving of slabs, in excess of 6 inches, is not uncommon as shown on Figure 6-1. Water marks and severe floor cracks indicate the extent of damage.

Figure 6-1. Differential slab heaving of up to 12 inches in a newly
completed basement.

2.   Broken utility lines often contribute water to the underslab soils.
     Water and sewer lines buried in expansive soils are subject to
     stress. Differential heaving can break pipes and cause leakage. Such
     leakage can continue for a long period of time without being
     detected.   In one case, the contractor neglected to connect the
     interior sewer line to the street sewer, and this fault went undetected
     until extensive damage had taken place. Figure 6-2 shows floor
     movement in a boiler room. The floor drain to the boiler room became
     plugged resulting in severe slab heaving from uplift.

3.   A most common source of moisture entering the underslab soils is
     derived from irrigation, lawn watering, and roof downspouts. Surface
     water enters the loose backfill and causes a wetting condition.

The above sources of water that enter the underslab soils are the
obvious ones. Moisture migration due to thermal differential as mentioned
in Chapter 2, can also cause damage to slab-on-ground without the observance
of free water.

Floor cracking caused by swelling soils must be differentiated from
that caused by shrinkage of concrete. In an expansive soil area, soil
heaving is unjustly blamed for all cracks that develop in a floor. Floor
cracks due to heaving generally take place along the bearing wall as shown
in Figure 6-1. In the absence of joints, shrinkage cracks can take place at
approximately equally spaced intervals. For concrete floors covering a
large area, the Portland Cement Association recommends the installation of
control joints at intervals of approximately 20 feet. Isolation joints
separating concrete slabs from columns, footings, or walls to permit both
horizontal movement due to volume changes and vertical movement caused by
differential settlement or heaving are also recommended.

Curling of concrete slabs in large floor areas due to improper curing
is not uncommon. Concrete curling has a strong resemblance to the uplift of
slabs due to the heaving of underslab soils.

Underslab gravel

Conventional slab construction uses 4 inches of gravel beneath all
concrete floors. A widely accepted theory pertaining to slab-on-ground
construction in expansive soil areas is that water from a single source,
such as from a broken pipe or from an improperly located downspout, travels

Figure 6-2.   Heaving of a floor slab in a boiler room. Source of water
              derived from inadequate floor drain system. Uplift is 2
              inches.

without resistance throughout the gravel bed and saturate the entire area
underneath the slab.   Therefore, more extensive damage to the floor takes
place when the gravel is used. To date, this theory has not been proven.

The use of gravel beneath the slab allows the uniform distribution of
floor load and the uniform curing of concrete, thus reducing the shrinkage
cracks and sometimes the curling of concrete. The main advantage of using
gravel beneath the floor slab, however, is to protect the building from the
rise of ground water.   If a perched water condition develops beneath a
basement which has neither a subdrainage system nor a gravel bed beneath the
slab, there is no easy method of removing the water. The installation of a
subdrainage system inside or outside of a completed building is a major
undertaking.   If, however, free draining gravel has been previously
installed beneath the slab, it may only be necessary to install a sump pump
in the basement as the water will flow through the gravel toward the sump.

In any event, the advantages of providing gravel beneath the slab far
exceeds any possible disadvantages.

STIFFENED SLABS

Slab-on-ground construction cannot be safely used in an area where the subsoil possesses high swell potential. For many years, both structural and soil engineers attempted to devise an economical floor system which would combat the problem of swelling soil. Unfortunately, such a system has not been devised to date. The systems now include the structural floor slab, the raised floor system, and the honeycomb system.

Structural floor slabs

The best method to prevent floor movement is to construct a structural slab supported on each side by grade beams and provide a void beneath the slab to prevent contact between the soil and the slab. The shortcomings of this system lie not only in the cost of construction, which is much more expensive than the conventional slab-on-ground method, but also upon the construction technique.

The use of a structural slab in an expansive soil is well accepted in Israel and South Africa, yet the resistance encountered in the Rocky Mountain Area for many years has been very strong, mainly from a cost consideration. Since residential construction is highly competitive, a builder doesn't want the risk of high costs to discourage the buyer. In view of the increasing number of slab-on-ground failures where the builder has to replace the basement floor two to three times, the added cost appears to be justified.

One distinct advantage of using a structural slab is the increase of dead load pressure on the foundation system, thus increasing the dead load pressure exerted on the piers as analyzed in Chapter 4.

The most convenient construction method is to provide a crawl space beneath the slab. This can be readily provided in major structures such as schools and office building. The crawl space provides access for inspection, can be ventilated, and can also serve as a convenient area for utility pipes and conduits. Either timber or concrete floors can be used in this type of construction.

Oftentimes, it is not possible to construct a crawl space, and the structural floor must then be constructed with only a few inches of air space between the slab and ground. This is typical where a structural slab is to be constructed in a basement area. The problem with this type of construction is that of providing a forming material to allow the placing of concrete. The use of Verticel, J-void, or other forming material similar to that used beneath the grade beams in the pier foundation is satisfactory. Forming materials are costly and there is no assurance that the material will completely deteriorate beneath the slab, thereby allowing the build-up of uplifting pressure. Commercial prestressed, hollow core, flat slabs are available in sufficient length to span 20 feet, thereby eliminating the need for void forming material. The use of prestressed slabs in large quantities can prove to be economical.

Raised floor system

The Portland Cement Association [6-1] has approached the problem of the construction of a structural floor slab on expansive soils by utilizing a concrete floor raised above grade by intersecting concrete ribs formed in a waffle pattern.

The raised floor system is constructed by placing Verticel or J-void (waxed cardboard boxes) upon a level subgrade. The spaces between the boxes contain reinforcing and form-supporting concrete ribs. The actual floor slab, containing wire fabric, is placed over the supporting ribs and cardboard boxes in a monolithic concrete placement. A typical plan and cross-section is shown on Figure 6-3.

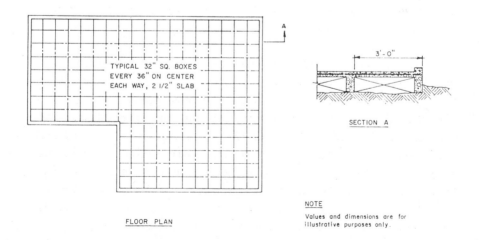

FLOOR PLAN

SECTION A

NOTE
Values and dimensions are for
illustrative purposes only.

Figure 6-3. Raised concrete floor system (after Portland
Cement Association).

The spacing of the ribs and the thickness of the slab depend upon the swelling potential of the surface soils and the anticipated dead load to be imposed on the waffle structure. The advantage of such a system is that it offers a clear, rational approach for the structural engineers, and the formed voids provide a means of relieving upward swelling pressure. The system can also incorporate utility routes, such as heating and cooling, through the floor. The disadvantage of such a floor system is the inability to exert sufficient dead load pressure upon the ribs to counteract the swelling pressure. This type of construction is also expensive. In addition, the floor area must be very finely graded to provide a level base for the void forming material so that uniform thicknesses result. This grading is an additional cost.

A recently developed cardboard forming system is shown on Figure 6-4. This system is cheaper in cost than the J-void system and is easy to install.

Honeycomb system

The development of the honeycomb system was based upon the assumption that comparatively slight movements of some clays reduce or relieve swelling pressures [6-3]. The foundation consists of longitudinally split Sonotubes that are placed with the openings toward the soil as shown on Figure 6-5, the bottom 2 inches of the space between Sonotubes being filled with sand. The Sonotube forms stand up well during placing of the concrete but disintegrate after being wetted. After the tube disintegrates, the sand runs out from under the joists.

It was theorized that as the clay swells, it expands into these openings and reduces the swelling pressure. The system has been tried in a few limited cases in the Denver area with doubtful success.

Both the raised floor system and the honeycomb system have the disadvantage of being too close to the ground surface. It is difficult if not impossible to ventilate the shallow space between the slab and the ground. Consequently, the development of dampness and mildew under the slab is likely to cause discomfort to the owner. With the development of a perched water condition, it is likely that the entire space becomes flooded and proper drainage in the shallow space is difficult.

Probably the most important disadvantage of such installations is that the systems cannot be inspected. In case problems develop beneath the floor, the only remedial measure is to tear up the slab. The author

believes that if a structural slab is to be installed, it is advisable to
provide at least a 3-foot-high crawl space so that proper ventilation and a
proper drainage system can be installed.   Also, inspection and any remedial
construction are easy.

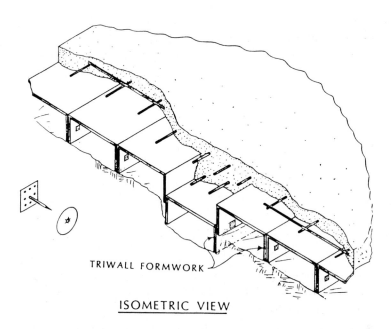

TRIWALL FORMWORK

ISOMETRIC VIEW

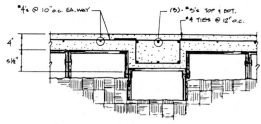

SECTION

Figure 6-4.   Triwall formwork.
(After KLP Consulting Engineers)

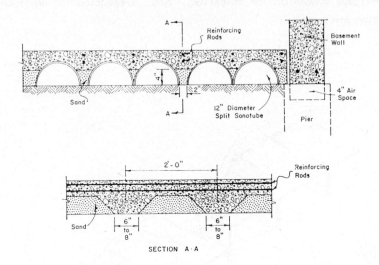

Figure 6-5. Typical honeycomb form system.

FLOATING SLABS

A floating slab refers to slab-on-ground construction in which the slabs are totally separated from the grade beam and the building structure. Theoretically, the slab is capable of moving independently without being in contact with the surrounding structure.

Unfortunately, there is no such construction as a truly floating slab. A true floating slab calls for the concrete slab to be totally free from the grade beams with no constraint exerted from the slab bearing members. As explained in the following sections, such construction is not likely to take place. Today, builders continue to advertise the floating slab system, the floors continue to exert pressure on the foundation walls, and engineers can do very little to relieve such a condition.

As long as the basement slab is used to resist earth pressure exerted on the basement wall, the concept of a "floating slab" cannot be realized.

Slip joints

Interior floor slabs should be totally separated from the grade beams and interior columns to allow for free slab movement. If the slab is not separated from the grade beam, heaving of the slab can transmit pressure to the grade beam and, in turn, lift the piers.

In practice, the slab is separated from the grade beams by the use of asphalt felt expansion joints. When the backfill exerts lateral pressure on the grade beam, the expansion joint is under compression and part of the uplift pressure beneath the slab is then transmitted to the grade beam. In nearly all basement buildings which have been subjected to uplifting, the central portion of the slab raised while the area along the perimeter of the grade beam remained essentially in place. In many cases, cracks appeared about 2 feet from and parallel to the grade beam, as shown on Figure 6-6.

Figure 6-6. Floor cracks parallel to the foundation wall resulting from
lack of slip joints.

Theoretically, if the swelling pressure of the underslab soil is 5,000 psf, as much as 10,000 pounds per linear foot of uplifting pressure can be transmitted to the grade beam. Naturally, as soon as the slab cracks, the uplifting pressure is relieved. However; the initial uplift force is sometimes sufficient to cause heavy damage.

An improved construction method is the installation of a lubricated slip joint between the grade beam and the slab as shown on Figure 6-7. This installation involves the use of two 1/8-inch masonite strips with a silicone lubricant between them. This type of joint system is not affected by lateral pressure, thereby allowing free slab movement.

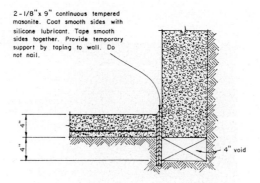

Figure 6-7.   Typical slip joint detail between slab-on-ground and foundation wall.

Some architects prefer to extend the floor slab into the exterior foundation wall as shown on Figure 6-8. The result is obvious; heaving of the floor slab not only produces floor cracks but also tilts the exterior wall, causing great structural damage. Figure 6-9 indicates the results of faulty design.

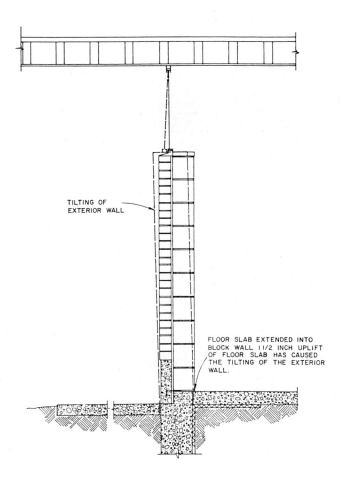

TILTING OF
EXTERIOR WALL

FLOOR SLAB EXTENDED INTO
BLOCK WALL 1 1/2 INCH UPLIFT
OF FLOOR SLAB HAS CAUSED
THE TILTING OF THE EXTERIOR
WALL.

Figure 6-8. Tilting of exterior wall caused by slab heaving and improper
            slip joints.

Exterior slabs

    Exterior patio slabs can also transmit swelling pressure to the
structure. Conventional practice calls for the exterior slab to be tied in
with the grade beam by the use of dowel bars. In this manner, full swelling
pressure is transmitted into the foundation walls. Such type design is not

Figure 6-9. Exterior brick course buckled due to heaving of interior
             floor slab.

recommended.   See also "Aprons" for a subsequent discussion of similar
problems.

     Figure 6-10 shows a typical case where the patio slab has transmitted
pressure through dowel bars to the foundation wall causing considerable
damage.

     In another case, the exterior sidewalk slab was extended about half an
inch into the brick course for aesthetic reasons (Figure 6-11).   This
resulted in the flaking and damaging of the brick wall, as shown on Figure
6-12.

     Oftentimes, the patio slabs are tied into the top of the foundation
wall with dowel bar as shown on Figure 6-13. Heaving of the patio slab can
cause severe cracking of the upper structure without any sign of movement
being apparent in the basement portion of the structure.

     A suggested method of constructing an exterior slab is shown on
Figure 6-14. In this case, the slab is totally free from the foundation

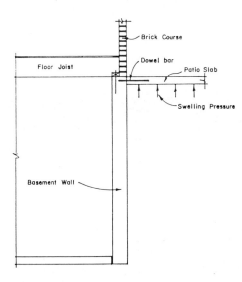

Figure 6-10. Patio slab dowelled into wall. Crack appeared parallel to
             wall. Magnitude of swelling pressure transmitted is estimated
             to be about 10,000 pounds per running foot.

wall and is free to move. The porch post is placed on the turned down slab
with an adjustable collar at the base. Such a post can be adjusted so that
the patio roof will not exert pressure to the house.

Partition wall

     The single largest cause of damage to structures founded on expansive
soils is partition walls that bear directly on a slab. When the floor slab
heaves, everything resting on the floor rises. The amount of floor heaving
depends upon the swelling potential of the underslab soils as well as the
degree of wetting. Slab heaving ranging from a fraction of an inch to as
much as 12 inches has been observed. The items affected by floor heaving
are:

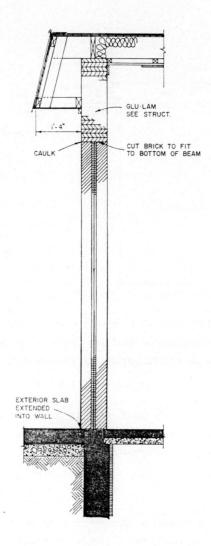

GLU·LAM
SEE STRUCT.

1'-4"

CUT BRICK TO FIT
TO BOTTOM OF BEAM

CAULK

EXTERIOR SLAB
EXTENDED
INTO WALL

Figure 6-11.   Flaking of brick course
caused by slab heaving.

Figure 6-12. Sidewalk heaving caused flaking of the brick course as a
direct result of the slab extending into the brick course.

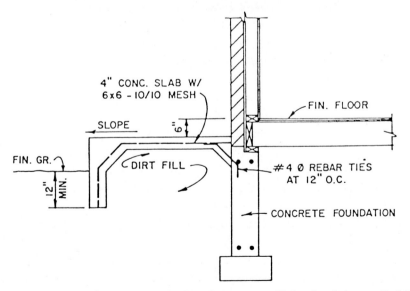

Figure 6-13. Patio slab attached to basement wall by dowel bar. Swelling
pressure results in damage to brick course.

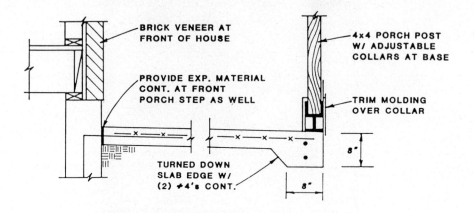

Figure 6-14. Suggested exterior slab construction detail.

Stud walls,

Sheet rock,

Wall paneling,

Cinder block partitions,

Staircase walls,

Door frames,

Water lines,

Furnace ducts, and

Shelves and bookcases

Any one, or several, of the above items can impart pressure to the upper floor joist or beams; initially, doors will bind followed by the occurrence of severe cracking. Almost every building investigated suffered some degree of damage due to the uplifting of the slab-bearing partition walls.

For stud wall construction, it is relatively easy to provide slip joints in the system so the wall is free to move without exerting pressure on the upper structures. A typical detail of such construction is shown on Figure 6-15. The slip joints can either be installed at the top (floor supported) or at the bottom (hung partition wall). The disadvantage of using a floor-supported wall is that when the wall lifts, vertical cracking occurs between the partition wall and the exterior walls as shown on Figure 6-16. Figure 6-17 indicates the architectural detail of a school building in which the load-bearing walls are supported by piers and interior masonry walls are placed on the slabs. Heaving of the slab-on-ground results in severe cracking of the slab-bearing partition wall while the walls supported by the grade beams and piers remain stable as shown on Figure 6-18.

Frequently, the stud walls are properly provided with slip joints; but sheet rock, applied on both sides of the studs, resting on the floor as shown on Figure 6-19 negates the slip joint installation. Sheet rock is capable of transmitting sufficient pressure to the floor joint or ceiling resulting in great damage.

A similar situation occurs when basement walls are paneled. The studs may be free from the floor, but the paneling bears directly on the floor. If uplifting occurs, it may result in the popping of the paneling as shown of Figure 6-20.

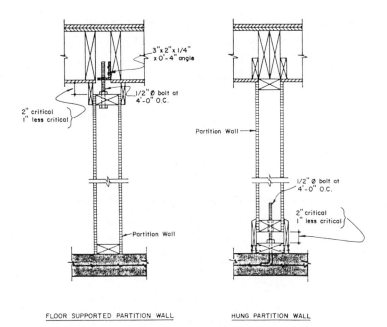

Figure 6-15. Detail of slip joints used in a partition wall. (After Jorgensen and Hendrickson, Inc.)

Figure 6-16. The cracking of slab bearing partition wall at the junction
of the exterior wall.

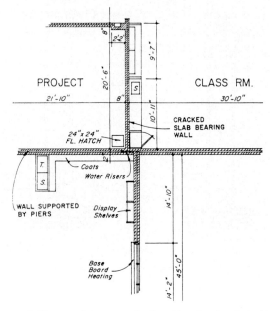

Figure 6-17. Architectural detail of school building.

Figure 6-19. Buckling of stud wall due to floor heaving.

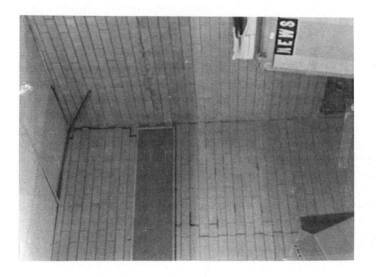

Figure 6-18. Heaving of interior slab-bearing partition. Note the uncracked wall, on the right, is supported by structural grade beams.

Figure 6-21. Staircase 2x4's rest directly on floor. Note bow of 2x4'[s due to floor heaving.

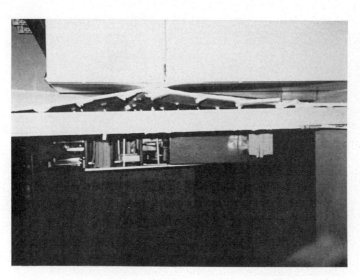

Figure 6-20. Buckling of wall paneling due to floor heaving

Staircase walls are the most frequently neglected when it comes to providing proper uplift precautions. When instructions are to provide slip joints to all slab bearing walls, the staircase is usually neglected. One single 2 x 4 can exert great uplift pressure and thus damage the upper structure, and the force may even extend to the second level in the case of split level buildings as shown on Figure 6-21.

Figure 6-22 indicates a properly formed slab-bearing partition wall with slip joints at the bottom.

Door frames and utilities

Door frames should be hung from the top and not supported on slabs. Slab heaving can transmit high intensity pressure through the door frame to the upper structures.

A very common distress in residential houses is the separation of garage door frames from masonry walls. This is essentially caused by heaving of the garage slab which results from failure to provide a grade beam across the entrance for the garage door opening.

Figure 6-22. Properly formed slab-bearing partition wall with slip joint
            at bottom.

Figure 6-23. Typical garage floor heaving at the central portion.

Figure 6-23 indicates a common sight of garage floor slab heaving in a swelling soil area. The central portion of the slab heaved; however, the edges remained in place because they were restrained by the door frame.

Figure 6-24 indicates the crushing and distortion of furnace ducts resulting from heaving of a basement slab. Figure 6-25 shows the severe bending of the water line also caused by floor heaving. Such distress generally brings immediate alarm to the home owner. If the utility lines above the slab are being damaged, those below the slab can also be seriously damaged.

Aprons

Concrete sidewalk slabs around the building prevent surface water from entering through the backfill and into the foundation soils. However, the concrete apron should not be doweled into the foundation wall for reasons previously discussed.

Concrete sidewalks or aprons heave and crack. Heaving of concrete walks can sometimes result in the drainage being directed toward the building, allowing surface water to enter through the joint between the apron and the wall. Frequent expansion joints are necessary to prevent

Figure 6-25. Distortion of water pipe due to floor heaving.

Figure 6-24. Crushing of furnace duct caused by heaving of basement floor slab.

Figure 6-27. Sidewalk slab heaving transmits pressure through slab-bearing timber frame to the building causing heavy structure cracking.

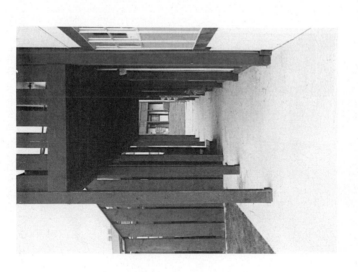

Figure 6-26. Typical heaving of sidewalk

excessive cracking of the slab.   Figure 6-26 shows the heaving of a concrete slab.

Another serious mistake is to place the posts of a patio on a concrete walk as shown on Figure 6-27.   As the concrete walk heaves, it transmits pressure to the structure through the connecting girders causing building damage.

REFERENCES

[6-1]   "Recommended Practice for Construction of Residential Concrete Floors on Expansive Soils," (Vol. II), Portland Cement Association, Los Angeles, California.

[6-2]   "Criteria for Selection & Design of Residential Slab-on-Ground", Building Research Advisory Board.

[6-3]   Means, R.E., "Buildings on Expansive Clay," Quarterly of the Colorado School of Mines, Vol. 54, No. 4, 1959.

Chapter 7

MOISTURE CONTROL

INTRODUCTION

Terzaghi stated that, "Without any water there would be no use for soil mechanics." Terzaghi had only limited knowledge of swelling soils; however, his statement can be accurately applied to the behavior of expansion soils. Ever since the acknowledgement of expansive soil problems, engineers have been attempting to isolate water from the foundation structure.

It is a relatively simple undertaking to remove free water which may seep into a building foundation by providing adequate surface drainage and properly installed subdrainage systems. However, it is difficult to isolate the migration of moisture from an exterior location to a covered area. Vapor barriers, both horizontal and vertical, have been used with only a limited degree of success in impeding moisture migration. Further research is necessary in both the field and the laboratory to establish a practical and economical method of controlling moisture migration.

HORIZONTAL MOISTURE BARRIERS

Horizontal moisture barriers can be installed around a building in the form of membranes, rigid paving, or flexible paving. The purpose of the horizontal barriers is to prevent excessive intake of surface moisture. The use and effectiveness of these moisture barriers are discussed below.

Membranes

A widely used horizontal moisture barrier is a combination of a polyethylene membrane extending beyond the limits of backfill and loose gravel placed on top of the membrane. The purpose of such installation is to prevent surface water from seeping through the backfill into the building and to prevent the growth of weeds. Figure 7-1 indicates the typical design.

It should be realized that the dry soils beneath an impervious membrane will, in time, become wet regardless of the presence of such a membrane because evaporation can no longer take place.

The thickness of the polyethylene membrane ranges from about 4 to
20 mils.   The membrane tears easily and eventually develops holes.   Surface
water ponding in a depressed area will in time leak through the holes and
edges of the membrane and enter the soil beneath, while evaporation and
drying of the soil beneath the membrane are impossible.   Even in the case of
a perfect impervious membrane, moisture migration due to thermal transfer as
explained in Chapter 2 will introduce additional moisture to the foundation
soils.

Thus, it appears that the questionable advantage of using a membrane
around the building is to increase the time required for moisture
penetration and make the moisture distribution more uniform.   In the course
of several years, the backfill soil beneath a membrane will be totally

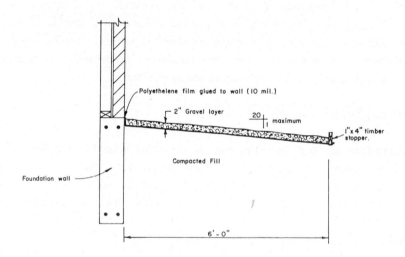

Figure 7-1.   Impervious membrane along exterior walls.

saturated.   By lifting the membrane, it is easy to find that the soil has a
moisture content greater than the plastic limit.

Concrete aprons

The installation of concrete aprons or sidewalks has been found
effective in controlling moisture fluctuation.   The advantage of using
concrete aprons rather than plastic membranes is that the former offers a

positive barrier to water. Obviously, within reason, the wider the concrete
apron the more protection it offers to the building. Paving the entire
non-building area is impractical and unsightly. Nonetheless, it has been
observed that foundation movement due to expansive soils seldom takes place
in gasoline service stations where the ground surface is extensively
covered.

Mohan and Rao [7-1] installed a four-foot wide concrete apron around
distressed buildings founded on black cotton soils, which proved effective
in controlling movement. They claim that the function of the apron is to
move the marginal moisture variation away from the building.

While the use of concrete aprons around the exterior of the building
may prove beneficial, care should be exercised in obtaining an effective
seal between the aprons and the foundation walls. Swelling soils can heave
an apron so that surface drainage is toward the building rather than away.
With poorly constructed joints, water will enter the joint and seep into the
foundation soil; thus an apron can cause more damage than good. In those
areas where concrete aprons are used, constant care and maintenance are
required.

Asphalt membranes

As early as 1933, the Texas Highway Department used asphalt membranes
to prevent surface water from entering the expansive clay subgrade [7-2].
The concrete pavement was placed upon an asphalt mat 48 feet wide. This
kind of membrane installation may retard swelling but will not prevent it.

Waterproofing membranes, particularly continuous sprayed asphalt such
as catalytically blown asphalt, have been used successfully in many state
highway departments. Asphalt membranes perform best when applied over the
entire subgrade as shown on Figure 7-2. Experience indicates that membranes
perform best where the soil profile is relatively dry; the moisture content
profile is relatively uniform with depth; the ground-water table is deep;
and the climate is dry to sub-humid. These conditions generally describe a
situation in which the major subgrade moisture is from surface infiltration
with the membranes eliminating the surface moisture ingression. Other
methods, such as full-depth asphalt pavement with sprayed asphalt or
synthetic fabric membranes beneath the verge slopes and ditches, provide a
useful treatment alternative [7-3].

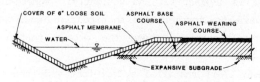

FULL-DEPTH ASPHALT PAVEMENT WITH LINED DITCHES

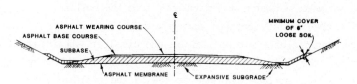

CONTINUOUS ASPHALT MEMBRANE APPLIED TO SUBGRADE AND DITCHES

Figure 7-2. Typical sprayed asphalt membrane applications to minimize
             subgrade moisture variations from surface infiltration.
             (After U.S. Dept. of Commerce).

Research conducted by the Asphalt Institute [7-4] advocates that asphalt membranes constructed from catalytically blown asphalt can be effective in preventing moisture from intruding into subgrade soils. Specifications for catalytically blown asphalt cement are given in Table 7-1.

Van London [7-5] used the 50-60 penetration asphalt membrane to completely envelop the highway embankment. The purpose of the membrane was to maintain a constant moisture content in the embankment soil, thus preventing volumetric change of the fill material. Subsequent moisture determinations of the enclosed embankment indicated very little change in moisture content over the 10-year period subsequent to construction; the pavement remained in a stable condition.

Another type of asphalt membrane consists of prefabricated asphalt sheets, less than one-half inch thick, three to four feet wide, and up to twenty feet long. Such material can be conveniently handled and easily placed.

Asphalt membranes can be used to cover the surface of expansive soils so that nonexpansive fill can be placed on top of the membranes. This will minimize the infiltration of surface water into the underslab soils. Where slab-on-ground construction is required, such treatment can be very advantageous. The amount of asphalt cement required to construct a membrane, according to the Asphalt Institute, is about 1.3 gallons per square yard.

The use of asphalt membranes in connection with the construction of a swimming pool in an expansive soil area is particularly desirable. Further research and field observation are required.

Table 7-1. Specification for catalytically-blown asphalt cement.

TEST DESIGNATION	TEST METHOD ASTM		50-60 PENETRATION GRADE
Flash point	D	92	425° F. Min.
Softening point	D	36	175° F.-225° F.
Penetration, 77° F.			
100 gms., 5 sec.	D	5	50-60
Penetration, 32° F.,			
200 gms., 60 sec.	D	5	30 Min.
Penetration, 115° F.,			
50 gms., 5 sec.	D	5	120 Max.
Ductility, 77° F.			
(5 cm per min) cm	D	113	3.5 Min.
Loss on heating			
325° F. in 5 hrs.	D	6	1.0 Max.
Penetration of residue,			
77° F. (100 gms., 5 sec.			
compared to original) percent			60.0 Min.
Solubility in CCl_4, percent	D	165	97.0 Min.

Asphalt used as a membrane shall be 50-60 penetration grade. This material shall be prepared by the catalytic blowing of petroleum asphalt. The use of Iron chlorides or compounds thereof will not be permitted. The asphalt shall be homogeneous, free of water and shall not foam when heated to 347° F. It shall meet the above tabulated requirements.

VERTICAL MOISTURE BARRIERS

Vertical moisture barriers are used around the perimeter of the building to cut off the source of water that may enter the underslab soils. Theoretically, vertical barriers should be more effective than horizontal barriers in minimizing seasonal drying and shrinking of the perimeter foundation soils, as well as maintaining long term uniform moisture conditions beneath the covered area.

Installation

Buried vertical barriers may consist of polyethylene membrane, concrete or other durable, impervious materials. The path of moisture migration when using a vertical barrier is shown on Figure 7-3.

As seen from Figure 7-3, the installation of a vertical barrier prevents edge wetting due to lateral moisture migration within the depth to which the membrane extends. However, over a period of time, rainfall, melting snow, and lawn irrigation water will accumulate near the bottom of the membrane and the moisture will be sucked into the moisture-deficient soils beneath the building. Thus, the same degree of wetting of the foundation soil can result with or without the use of a barrier but would occur over a longer period of time. By installing a moisture barrier, the potential for damage is less because of the slower rate of moisture migration and the more uniform moisture content of the soil at any particular time.

Vertical moisture barriers should be installed to a depth equal to or greater than the depth of seasonal moisture changes.

A study in South Africa [7-6] has shown that the wetting of soils beneath a house occurred to a depth of at least 24 feet in a fissured-clay profile. Most of this wetting was attributed to lateral migration of moisture from seasonal rains, rather than from capillary rise.

The following should be considered in the installation of a vertical moisture barrier:
1. Vertical moisture barriers cannot be effectively installed around basement structures.
2. Vertical moisture barriers should be installed two to three feet from the perimeter foundation to permit machine excavation of the trench for the membrane.

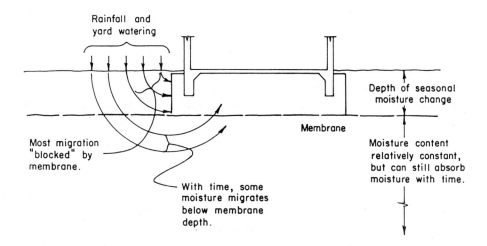

Figure 7-3. Path of moisture migration blocked by vertical barrier.
(After Woodward-Clyde Sherard and Associates).

3. The vertical barrier is sometimes attached to a horizontal barrier to
    prevent wetting between the vertical barrier and the building.
4. Either concrete or polyethylene membrane can be used. The membrane
    should be of sufficient thickness and durability to resist punctures
    during backfilling of the trench.
5. It is also possible to use semi-hardening, impervious slurries
    installed in a narrow trench.

Theoretically, vertical moisture barriers have a distinct advantage
over horizontal moisture barriers. However, in view of the high cost
involved in the installation of a vertical moisture barrier, especially
where great depth is required, it is doubtful that such an installation is
of sufficient merit to warrant the expense.

Deep vertical seal

Probably the most successful and well documented use of moisture
barriers to control pavement distress is that of the experiment in San
Antonio, Texas. Since horizontal moisture seals over subgrade cannot be

applied to existing highways, the use of a vertical moisture barrier
appeared to be a logical solution.

The first Texas highway test section is on IH-410 in southwestern San
Antonio [7-7]. The soils are the Houston black and Houston Association,
deep calcareous and gravelly clays. They have a very high shrink-swell
ratio. Their plasticity index ranges from 35 to 50, and the liquid limit
from 56 to 72.

The existing highway was built in 1960. Since its construction,
subgrade activities have been reflected in repeated asphalt level-ups of the
pavement, continuous surface distortions and irregularity in the curb
profiles.

The fabric test section was placed through a zone of activity at the
edges of the northbound lanes outside and inside shoulders. The trench was
three feet wide and nine feet deep, as shown on Figure 7-4. The fabric
consists of DuPont Typar spun bonded Polypropylene with EVA coating.

Following fabric placement, moisture sensors were placed inside and
outside of the protected area, as well as under the adjacent southbound
lane. The first reading took place in May, 1979. In November, 1979,
fourteen of the forty-six sensors indicated resistance reflecting a drying
condition. Three of six sensors under the southbound control lane indicated
drying. Eleven of the remaining forty sensors showed drying conditions
along the northbound lane. Only four of the eleven were inside the fabric
protected area. This initially indicated that the fabric-protected subgrade
was less likely to have moisture changes that could result in pavement
elevation changes.

Vertical membrane cutoffs have not been used extensively primarily
because of construction problems. Ideally, the depth of the vertical
membrane should extend to the depth of the active zone. To date, the merits
of such installation on a long-term basis have not been evaluated.

Backfill

An important element involved in building construction which is usually
slighted is the backfill around a building. When properly placed and
compacted, backfill serves the same purpose as the vertical moisture
barriers. This is especially true in basement construction where properly
compacted backfill can prevent surface water from entering the foundation
soils.

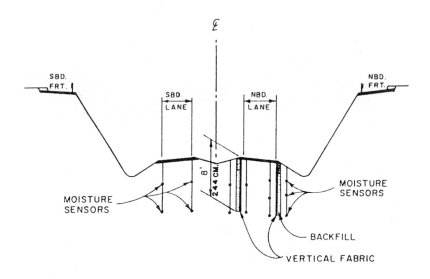

Figure 7-4. Interstate 410 section (after M.L. Steinberg).

Unfortunately, backfill is seldom compacted properly. Many builders choose to push loose soil into the excavation with no further compaction efforts. Others attempt to consolidate the backfill by puddling. It is obvious that proper compaction of the backfill cannot be achieved by such processes. When improperly compacted, almost all of the backfill along the foundation walls is in a loose state. Surface water can then enter the backfill and seep freely into the foundation soils. The result of settlement of loose backfill is shown on Figure 7-5.

Compaction of backfill in restricted areas, such as in utility trenches, cannot be performed by large compacting equipment. These areas should be compacted, by hand-operated vibrating plates, in thin lifts of not more than four inches. Because of poor compactive efforts, the settlement of backfill around a building as well as the settlement of utility trenches is the rule, rather than the exception.

One recently advertised hydraulic-operated compactor claimed to have the ability to compact backfill in a trench, in one effort, from the ground surface. Such a claim appears false while, in fact, the compactor has probably a lift compaction capacity of no more than 12 inches.

Figure 7-5. Typical case of loose backfill around the building. Loose
             backfill allows surface water to enter the foundation soils.
             Note depressions which trap water.

The so-called puddling process has been widely used by contractors in
small jobs with the assumption that the soils will consolidate without
compaction.   Not only does such practice make it impossible to obtain the
required density but it is also sometimes dangerous.   In cohesive backfills,
puddling inevitably leads to weakening and softening of the soil and to
future loss of stability and subsidence.   In uniformly granular soils,
puddling will cause the collapse of the extremely loose, unstable zone
associated with bulking.   Excess puddling, which frequently results when a
hose is left discharging into the backfill overnight, can easily introduce
water into the foundation soils and beneath the slab, resulting in great
damage.   In expansive soil areas, such damage will be reflected in the
structure for many years.

By using impervious clay compacted to 85 percent or more of standard
Proctor density at optimum moisture content, the backfill acts as a very
effective vertical moisture barrier.   Such barriers are more effective than
membranes.

SUBSURFACE DRAINAGE

The purposes of a subsurface drainage system are as follows:
1. Intercept the gravity flow of free water,
2. Lower the ground water or perched water, and
3. Arrest the capillary moisture movement and movement of moisture in the
    vapor state.

Intercepting drains

Intercepting drains are effective in minimizing the wetting of the
foundation soils where the wetting is due to the gravity flow of free water
in a subsurface pervious layer such as a layer of gravel or fissured clay.
This is shown in Figure 7-6. To ensure the interception of free water, the
drain must be completely filled with gravel and the trench should be deep
enough to reach the water-bearing layer.

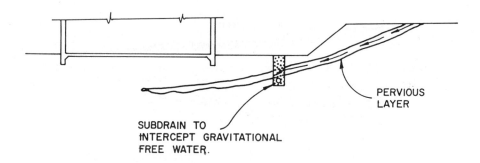

Figure 7-6. Typical function of an intercepting drain. (After
Woodward-Clyde-sherard & Associates).

Intercepting drains are most effective when located along the toe of a
slope where ground water leaves the deep strata and where it may emerge to
the surface. When a structure is located near an irrigation ditch or canal
with a leakage problem, the installation of an intercepting drain will
protect against the infiltration of seepage water. Intercepting drains are
also widely used for improving slope stability and preventing landslides.

Perched water

     According to textbook definitions [7-8], a perched water table occurs
when water seeping downward is blocked by an impermeable layer of clay or
silt and saturates the area above it, as shown in Figure 7-7.

     Water that infiltrates the soil in the uplands gradually moves
downward, eventually becoming trapped beneath an upper confining bed of
clay. Depending on the elevation of the water source, the permeability and
rate of flow in the aquifer, the rate of discharge and other factors, the
pressure in a confined aquifer will vary considerably.

     A majority of bedrock in the Rocky Mountain region consists of
claystone shale. It is Morrison, Mancos, Pierre or other formations. Some
of the claystone shale is interlayered with thin sandstone lenses which are
water bearing. The development of a perched water condition can be
summarized as follows:

1.   The upper soils are relatively pervious and surface water is capable of
     seeping through the upper soils encountering relatively low resistance.

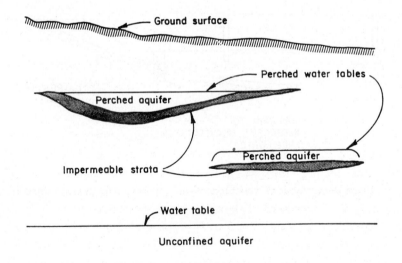

Figure 7-7. Perched aquifers. (After Todd).

2.   The lower bedrock is impervious and will not allow the infiltration of
     water. However, the deposit contains seams and fissures in erratic

directions. The thickness of the lenses varies from 1/16-inch to more than 1/4-inch. The spacing of these seams can be very short, only a few inches to very wide spacing of more than ten feet. The seams can be a closed system with no connections and can be continuous, leading to exterior water sources.

3. Claystone shale can be highly fractured, especially at the outcrop area, and can allow the penetration of water. At depth, however, seams and fractures are much less frequent, due to high preconsolidation stress.

4. With surface irrigation and precipitation, the surface water tends to seep through the upper soil and accumulate on top of bedrock. Part of the accumulated water will flow on top of the bedrock and part of it will run through the fissures of the bedrock. When a deep basement is constructed, the basement excavation can cut through the fissures of bedrock and the water can accumulate in the low basement area.

5. Perched water follows the contour of bedrock and does not necessarily follow the contour of the ground surface. Consequently, it is possible for a house sitting on the top of a nob to experience water problems while a house located in a low-lying area to be free of seepage.

6. The overburden of a claystone shale area where problems occur is generally thin, from soil exposed at the ground surface to a depth of about ten feet.

7. Perched water is fed by surface water derived from precipitation, snow melting, lawn irrigation, breakage of water mains, and other sources. The source of water need not be located immediately adjacent to the perched water reservoir. The reservoir can be replenished by a water source as far as one mile away.

8. The size of a perched water reservoir can vary considerably. A small perched water reservoir can pose seepage problems only after a prolonged wet season, while some perched water reservoirs do not dry up even during dry months.

9. Fluctuation of perched water directly reflects the seepage of water into a basement. Oftentimes, perched water can rise by a flash storm or heavy irrigation immediately adjacent to the property.

10. Wells tapping perched aquifers yield only temporary or small amounts of water. However, such water can enter basements and cause considerable damage. The installation of a subsurface drainage system around the perimeter of the lower level of a structure can protect against infiltration from perched water.

Mechanics of flow

The mechanics of water flow in the claystone shale is not fully
understood. The location, thickness and path of the seams and fissures in
claystone bedrock are so complicated that it is not possible to apply
Darcy's Law or any other mathematical relationship to obtain a rational
answer. Two systems of flow are possible.

The first case is one where the seams and fissures are not continuous
and not connected with an exterior water source. In such cases, water will
not be present in the seams until the entire claystone mass is saturated
since the permeability of the claystone mass is generally so low that the
development of a perched water condition takes a long time.

The second case is one in which the seams and fissures are continuous
and connected to an outside source. In many cases, it is the upper highly
fractured claystone. Water flows through the tortuous winding path and
eventually reaches an outlet. The flow of such water may not follow the
gravitational force, with capillarity playing a dominant role. Before the
subdivision development or the excavation has started, water from the
exterior source that feeds into the seams is limited and a perched water
condition seldom exists. During the drilling of the piers, it is uncommon
that water is found in the holes, unless ground water is encountered.

A perched water table condition usually develops after the following
sequence of events:
1. A completed subdivision has been occupied for two years or more.
2. Individual lawns are completed and extensive watering is practiced.
3. The streets and lanes are covered with pavement and the evapotrans-
    piration balance is disturbed.
4. There is an undetected leakage in the water or sewerline.
5. The backfill around the house is loose and surface drainage is poor.

When one or more of these conditions develop, the surface water is able
to enter the seams and fissures of the claystone, filling the space.
Gradually, even with low permeability, water is able to saturate the clay
between the seams. At the onset, the area immediately adjacent to the seams
develops a positive pressure. This can be easily identified by the water
level in the piezometer inserted through the seams. A negation pressure
will develop in the area where seams and fissures are not present, as
identified by the absence of water in the piezometer [7-8].

This condition may continue for a long time, depending on the amount of
surface water allowed to feed into the system. Eventually, where bedrock is

near the ground surface, a perched water condition prevails.

Hydrologists and ground water geologists may not agree with the above mechanics of flow. We are sorely in need of extensive field tests and large-scale laboratory testing so as to better understand this condition.

PERIPHERAL DRAINS

The purpose of a peripheral drainage system is to prevent free water from entering the lower floor of the building. In order to achieve this purpose, it is necessary that the drain be installed with adequate depth, capacity and proper outlet. A poorly designed drain system not only cannot intercept seepage water, but also it tends to hold water and create problems of slab swelling in an expansive soil area.

In the past, little attention has been given to the detailed design of the subdrain system. As-built drawings often only show a dotted line along the foundation wall with no section, gradient or outlet specified. As a result, a great majority of drain systems do not function. This has been especially true in the case of single-family homes where occupants are bothered by the wet basement problem.

Source of water

Water may seep into the building from the following sources:
1. Surface water collected around the building seeping through the loose backfill and eventually entering into the lower floor.
2. Perched water table that has been created after the area has developed.
3. Rise of ground water.
4. Long-term moisture transfer from high temperature area to the covered low temperature area or capillary rise.

The peripheral drain is effective in minimizing general wetting of the foundation soils, which occur not only because of gravitational flow of free water, but also because of moisture migration [7-9]. As explained in Chapter 2, moisture migration includes capillary moisture movement in the liquid state and movement of moisture in the vapor state due to temperature differential. Where the water table is deep, capillary action and vapor transfer are probably the major causes of wetting of the moisture-deficient soils in a covered area.

It should be noted that, for a peripheral drain to be effective in preventing moisture movement discussed above, it must be designed as a capillary break; and additionally, the vapor pressure in the drain should be at a lower value than the vapor pressure in the foundation soils. The gravel used to fill the subdrain trench should have a gradation between 3/4 and two inches in size with a percent of fines less than five.

Positive outlets should be provided for the subdrainage system. If a gravity system is not possible, the drain should be discharged to a sump where water can be removed by pumping. In some instances, it may be permissible to connect the subdrain to the gravel bed beneath the street sewer line. However, because of the small gradient differential between the sewer line intake and the subdrain outlet, such an arrangement is usually unsatisfactory.

Experience indicates that to be fully effective, the peripheral drain should be placed at least 12 inches below the floor level, preferably 24 inches.

A common opinion is that if free water can be prevented from entering the foundation soil, swelling of underslab soils does not take place. This is not necessarily true. Increase of moisture content without the presence of free water can easily cause severe heaving of the floor slab.

Moisture increase can be caused by moisture transfer from a high temperature area to a low temperature area. It can also be caused by moisture flow from soils of low clay content to those of high clay content. The difference in moisture content will cause flow from soils of high moisture content to soils of low moisture content [7-10].

The subdrain system is effective in controlling the water table so that water does not seep into the basement. However, capillary action is capable of drawing moisture from the controlled water level and cause swelling. Theoretically, the height of capillary rise for tight clay can be several hundred feet. In practice, little effect will be experienced if the water table is kept at a depth of ten feet below the floor level.

With the above reasoning in mind, it is clear that it a is false hope that basement floor heaving can be arrested by the installation of a subdrain system.

Subdrain design

Installed inside the basement. This type of installation is mainly used for the purpose of controlling the rise of the water table.

In the area where a perched water condition prevails or where there is a potential risk of the rise of ground water, a subdrain system as shown in

Figure 7-8 probably is the most effective. This system is not limited in disposing of surface water that may seep into the backfill.

By placing the drain inside the basement, it can be effectively connected with the underslab gravel, thus providing an effective drawdown of the water table. The drain should be installed deep enough so as to ensure the interception of the water table.

The main disadvantage of such a system is that it is necessary to provide a trench around the inside perimeter of the basement. Such an excavation may have to be accomplished by hand, which increases the cost of the drain system considerably. However, if the contractor is aware of such a necessity, he can provide the trench during the excavation for the basement.

Considering that if water seeps into the basement, it will cost at least several thousand dollars to remove the slab, dig the trench and install the system, the initial installation as a precautionary measure may be well worthwhile.

Installed outside of basement. Figure 7-9 shows the recommended installation of a subdrain for general conditions. It has the advantage of controlling the rise of the water table, seepage of surface water, as well as controlling hydrostatic pressure exerted on the backfill. However, where rise of ground water or development of a perched water table condition is critical, such a system is not as effective as that of a subdrain installed inside the basement. This is because the drain system is not directly connected with the underslab gravel. Connection of the drain system with the underslab gravel can be accomplished to some extent by means of the void space beneath the grade beam, but careful construction is required. It is even more difficult to connect to the underslab gravel if continuous wall footings are used as the foundation.

A properly installed subdrain system with the depth of drain placed at least three feet below the top of the floor slab can substantially decrease the potential floor heaving; and if floor heaving does take place, the movement will be more uniform.

On the other hand, when a slab is badly damaged from heaving caused by rise of perched water, subsequently the floor slab is removed and a subdrain properly installed. Such remedial construction does not arrest future slab heaving. Moisture in the soil does not decrease on account of the drain system. Continuous swelling of the underslab soil is to be expected.

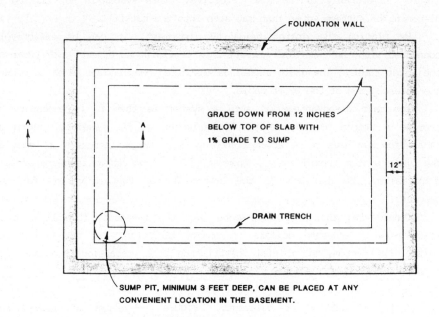

FOUNDATION WALL

GRADE DOWN FROM 12 INCHES
BELOW TOP OF SLAB WITH
1% GRADE TO SUMP

12"

DRAIN TRENCH

A

A

SUMP PIT, MINIMUM 3 FEET DEEP, CAN BE PLACED AT ANY
CONVENIENT LOCATION IN THE BASEMENT.

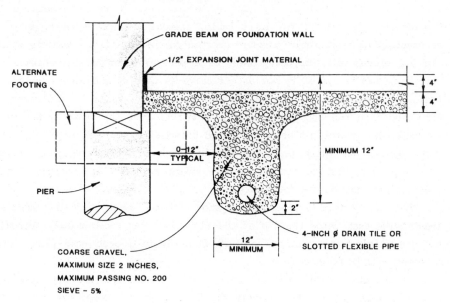

GRADE BEAM OR FOUNDATION WALL

1/2" EXPANSION JOINT MATERIAL

ALTERNATE
FOOTING

4"

4"

0-12"
TYPICAL

MINIMUM 12"

PIER

2"

COARSE GRAVEL,
MAXIMUM SIZE 2 INCHES,
MAXIMUM PASSING NO. 200
SIEVE - 5%

12"
MINIMUM

4-INCH Ø DRAIN TILE OR
SLOTTED FLEXIBLE PIPE

SECTION A-A

Figure 7-8. Typical subdrain detail located inside of basement.

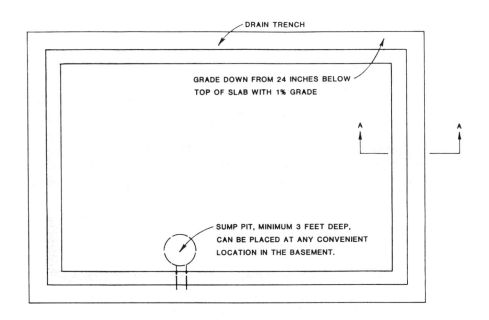

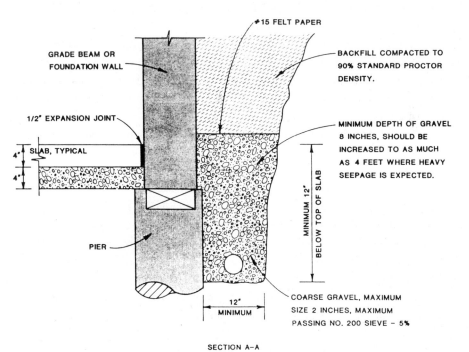

SECTION A-A

Figure 7-9. Typical subdrain detail located outside of basement.

This leads to the conclusion that the purpose of a subdrain system is to prevent the infiltration of free water into the basement. Any advantage the drain system can contribute toward swelling of the underslab soil is limited to providing more uniform movement.

Deep drain

The immediate reaction to dealing with the problem of perched water is the installation of a subdrain system. It remains to determine the capability of a subdrain system to drain water from the seams and fissures of the bedrock. It is necessary to evaluate the effect of a drain system on pier uplift and on slab heaving separately.

Subdrains installed at a shallow depth have little or no effect on pier uplift. At best, such drains will remove some of the water present in the upper few feet of bedrock. Even this is questionable. Theory indicates that unless the following two criteria are met, the drain is not effective:

1.   The seams and fissures in the claystone are connected to the exterior sources of water with the seams and fissures continuous.
2.   The entire soil mass is saturated.

In order to prevent perched water from reaching the perimeter of the pier and causing uplift, it is necessary to intercept the flow before reaching the pier. The installation of a deep drain system as shown on Figure 7-10 may intercept the perched water before reaching the piers. The following assumptions are made:

1.   The seams and fissures in claystone are connected to a source of water located at the exterior of the building.
2.   The claystone bedrock within the boundary of the building remains dry.
3.   The seams and fissures of bedrock are continuous.

This concept was used for the El Paso Community College in Colorado Springs at great expense. The claystone shale at the project site has a swelling pressure in excess of 50,000 psf. The building is now six years old with no apparent distress. It is too early to evaluate the performance of such a drain system, nor is it possible to justify the cost of such installation. It is obvious that for an average residential house, such an elaborate drain system is prohibitive in cost.

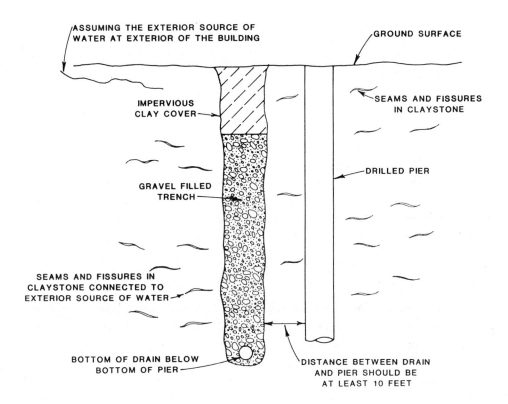

Figure 7-10. Deep drain system.

SURFACE DRAINAGE

The ground surface around a building should be graded so that surface water will drain away from the structure in all directions. This usually is not accomplished due to negligence, cost, limited property size and other reasons. In many cases, the area around the building has been properly graded after construction, but the grade was later changed to improve the appearance of the landscape. As a result, it is not uncommon to find buildings with surface drainage directed toward the foundation walls. Moisture change at the perimeter of the building appears to be the most significant contributor to damage. Therefore, by improving the drainage, a beneficial effect is inevitable.

Sprinkling system

Lawn sprinkling systems often create foundation soil problems. These systems should be installed at least ten feet from the building. Nozzles of the sprinkling system should never be directed toward a building. An automatic timing device should be provided for all sprinkling systems so that excessive watering is avoided. Before use, all outlets in the system should be pressure checked to detect the presence of any possible open outlet underground where water could flow unchecked for a long period of time without arousing suspicion.

Vegetation

From an architectural standpoint, it is pleasing to have shrubs and flower beds planted adjacent to buildings. However, since it is necessary to irrigate flower beds and shrubs, the excess water penetrates through the loose backfill into the foundation soils. Experience indicates that in practically every investigation of a cracked building, shrubs and flower beds are located adjacent to the building. Figure 7-11 indicates a typical case where drainage away from the building is obstructed by the paved walk and improper compaction results in a depression along the building, creating a ponding condition. Shrubs planted along a wall is typical of many school buildings and residential houses.

Many studies have shown that large bushes and trees can cause differential drying [7-11] of the foundation soils and result in damage to the building from shrinkage. Most of the damage caused by shrinkage takes place in non-swelling or low-swelling soil areas. It is doubtful whether large trees will pose a problem in high-swelling soil areas. Nevertheless, it is good practice to plant trees and shrubs at least ten feet from a structure.

Figure 7-11. Plantation strip around a school building allowing the
              ponding of water between the sidewalk and the foundation
              wall.

Roof drain

     Roof downspouts must be directed away from a structure so that water
does not seep into the foundation soils. The downspouts should extend well
beyond the perimeter of the foundation and should discharge to an area where
the surface drainage is adequate to carry off the water rapidly and prevent
any possible ponding of water. If necessary, the water from downspouts
should be carried in a closed pipe or lined ditch to the street. Many
modern buildings are constructed without downspouts, and water from the roof
drains freely through the loose backfill into the foundation soils.
     Sometimes an open courtyard is constructed in the central portion of a
building. The courtyard is usually covered with lawn, flower beds, and
trees. Such a courtyard constitutes a major drainage problem because

surface water will be unable to drain unless an adequate subsurface drainage
system is provided.

Interior plumbing

Interior plumbing, including sewer and water lines, should be carefully
checked for leakage. Sewer lines laid beneath the basement are subjected to
stress when the surrounding soils expand, and in extreme cases, shearing
stress causes pipe breakage resulting in flooding.

In one instance, while investigating a cracked house, it was found that
the plug on the drain trap beneath the shower stall was missing. As a
result, all the water from the shower drained into the crawl space area for
a period of at least three years.

Leakage from water lines is less frequent. Sometimes leakage is found
near the water meter which generally is situated near the sidewalk in front
of the house. Leaking water follows the loose backfill around the water
pipe into the foundation soils, causing damage.

REFERENCES

[7-1]   Mohan, D. and Rao, B.G., "Moisture Variation and Performance of
         Foundations in Black Cotton Soils in India," Moisture Equilibria and
         Moisture Change in Soils Beneath Covered Areas, Australia,
         Butterworth.

[7-2]   "Report of Committee on Warping of Concrete Pavements," Highway
         Research board Proceedings, Vol. 25, 1945.

[7-3]   "Technical Guidelines for Expansive Soils in Highway Subgrades,"
         U.S. Department of Commerce, National Technical Service, PB80-
         139660, June, 1979.

[7-4]   "Asphalt Membranes and Expansive Soils," Information Service No.
         145(IS-145) May 1968.

[7-5]   Van London, W.J., "Waterproofing Value of Asphalt Membranes in Earth
         Fills for Gulf Freeway," Proceedings, Association of Asphalt Paving
         Technologists, Vol. 22, 1953.

[7-6]   Collins, L.E., "Some Observations on the Movement of Buildings on
         Soils in Vereening and Odendaalsrus," Symposium on Expansive Clays,
         South African Institution of Civil Engineers.

[7-7]   Steinberg, Malcolm L., "Deep Vertical Fabric Moisture Seals,"
         Transportation Research Board, 1986.

[7-8]   Todd, D.K., "Ground Water Hydrology," John Wiley & Sons, Inc., 1959.

[7-9]   "Remedial Methods Applied to Houses Damaged by High Volume Change
         Soils," Woodward-Clyde-Sherard & Associates, Oakland, California,
         1968.

[7-10]   Haliburton, Allan and Dan Mark III, "Subgrade Moisture Variations in
         Expansive Soils," Transportation Research Board, 1986.

[7-11]   Hammer, M.J. and Thompson, O.B., "Foundation Clay Shrinkage Caused
         by Large Trees," Journal ASCE, Soil Mechanics and Foundation
         Division, Vol. 92, No. SM 6, Nov. 1966.

Chapter 8

SOIL STABILIZATION

INTRODUCTION

Expansive soils are costing the United States an estimated four billion dollars a year as stated in Chapter 1. More than half of these damages occur to transportation facilities. Highways, railroads, runways, canals, bike paths and pedestrian walkways are all among the casualties.

Unlike structure foundation, dead-load pressure cannot be utilized to limit the swelling of pavement and non-load-bearing structures.

In theory, the swelling potential of an expansive clay can be minimized or completely eliminated by one of the following methods:

1. Flood the in-place soil to achieve swelling prior to construction,
2. Decrease the density of the soil by compaction control,
3. Replace the swelling soils with nonswelling soils,
4. Change the properties of expansive soils by chemical injection, or
5. Isolate the soil so there will be no moisture change.

Isolation of the soil has been extensively discussed in Chapter 7.

Special emphasis will be directed in this chapter to highway pavement construction. It is obvious that the stabilization of the highway subgrade against swelling soils deserves immediate attention.

PREWETTING

An old established concept among engineers and contractors as well as laymen in dealing with swelling soils is prewetting. As explained in Chapter 2, moisture can migrate from a moderate-depth water table to an upper moisture-deficient soil by means of capillary rise. Moisture migration can also take place from a high-temperature area to a low-temperature area by means of thermo-osmosis or other mechanisms. Normally this moisture evaporates at the surface and moisture equilibrium is maintained in the soil. The presence of covered areas, such as floor slabs, pavements, or similar structures which inhibit this evaporation, increases the moisture content of the foundation soil with resultant swell.

The prewetting theory is based on the assumption that if soil is allowed to swell by wetting prior to construction and if the high soil moisture content is maintained, the soil volume will remain essentially constant, achieving a no-heave state; therefore, structural damage will not occur.

Ponding

The present prewetting practice usually involves direct flooding or ponding of the building area. The foundation and floor area are flooded by constructing a small earth berm around the outside of the foundation trenches to impound the water. Another practice includes first prewetting the foundation trenches, then placing the foundation which is used as a dike to flood the floor area. In some cases, where the moisture content at the footing depth is stable, it is possible to place concrete footings and utilize them as dikes so that only the floor area is prewetted.

The effect of ponding or flooding on the moisture content at various depths has been investigated by the Texas Highway Department [8-1]. A section of Interstate Route 35 north of Waco, Texas was chosen for the experiment. The subgrade was ponded and the moisture content at various depths was taken. The moisture variation at specific depths beneath the ponding area is shown on Figure 8-1. The following observations were made:

1.   The moisture content achieved a significant penetration of only four feet below the pond during a period of 24 days.
2.   To obtain desirable moisture distribution at greater depths, ponding should extend approximately 30 days.

Experience with highly expansive Taylor Marl east of Austin, Texas is presented by John Stevens and Hudson Matlock [8-2]. The soil consists of calcium montmorillonite with a liquid limit of 80 and a plasticity index of 22 to a depth of approximately 50 feet. The site was ponded with water for a period ranging from one week to 26 weeks. It is obvious that water is able to penetrate to a greater depth for soils with high permeability. Figure 8-2 shows the changes in moisture content profiles during ponding for 26 weeks for one value of saturated hydraulic conductivity. The depth of significant moisture penetration is less than two feet. Such penetration is considerably less than the Waco experiment.

Experience in Southern California [8-3] indicates that prewetting moderately expansive soils to a condition of 85 percent saturation at a depth of 2 1/2 feet is often satisfactory. In the case of highly expansive soils, prewetting to as much as three feet may not be sufficient.

For slab-on-ground construction, after completion of the prewetting
treatment, the ground surface must be kept moist until the slab is placed.
A gravel or sand bed four to six inches thick should be placed over the
subgrade prior to the prewetting period. The gravel layer prevents the clay
from drying and shrinking.

The prewetting operation must not be at the discretion of a contractor
or owner. The treatment should be based upon an engineering investigation
and evaluation of the site, subsoil condition, swelling potential, climatic
condition, foundation system, and prior local experience. The moisture
content profile should be checked frequently by tests in the field to assure
that the desired results are achieved.

Practice

Ponding or sprinkling, to increase the soil moisture to a degree that
will prevent harmful heaving upon subsequent wetting, has been used in the

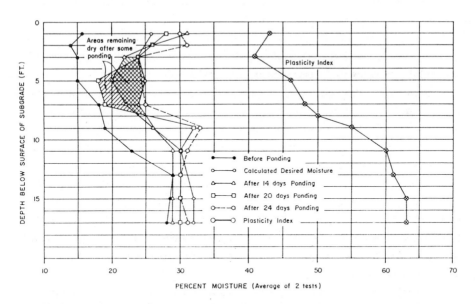

Figure 8-1. Subgrade moisture movement on IH-35. (After McDowell)
McLennan Co., Texas.

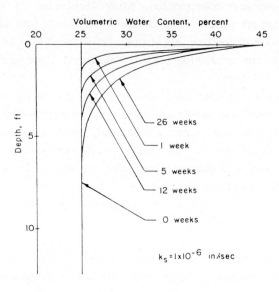

Figure 8-2. Water content profile under a pond (After Stevens and Matlock).

construction of the San Luis Drain on the San Luis Unit of the Bureau of Reclamation's Central Valley project in California.

Bara [8-4] claimed that if the dense clays with a particular liquid limit could be expanded to densities at or above the critical natural density-liquid limit reference line, a stable or near ultimate moisture condition would have been approached and future volume changes would be small. The liquid limit versus dry density relationship is shown on Figure 8-3. A soil with a liquid limit of 70 intercepts the reference line at 90 pcf density. Similarly, the water content-liquid limit relationship was developed for a soil liquid limit ranging between 40 and 100 as shown on Figure 8-4. Moisture contents above the reference line in Figure 8-4 would assure that the densities were on the non-critical side of the reference line in Figure 8-3.

Figure 8-4 indicates that clays at a liquid limit of 40 require only about 23 percent moisture, while those near a liquid limit of 100 require at least 37 percent moisture before they are considered to be relatively nonexpansive in situ.

Large scale experiments of flooding of foundation soil for building sites have been conducted in Vereeniging, South Africa [8-5]. Here, the effect of wetting was accelerated by a grid of vertical four-inch-diameter wells each 20 feet deep. At the end of 96 days, over 90 percent of the maximum surface had taken place. It is concluded by the authors that the acceleration of heave by flooding is a feasible pre-construction procedure for light structures.

E.J. Felt [8-6] discusses a prewetting project in which the soil moisture content did not increase appreciably after the first month of prewetting. For five months thereafter, soil swelling continued. It was suggested that the first infiltration of water was probably taken by seams and fissures present in the clay and, therefore, full soil expansion did not occur. As time passed, the water moved from the fissures into the blocky soil mass, and swelling took place throughout the mass of the soil, not merely along a seepage path.

At a housing project near Austin, Texas, the expansive soil beneath the foundation was prewetted by filling the foundation trench with water. After six weeks of soaking, the water was pumped out, the foundation placed on the wet soil, the trenches again filled with water, and the soil kept wetted thereafter. This house heaved both during and after construction.

It was concluded by Dawson [8-7] that it is extremely difficult to saturate high plasticity clays within a reasonable period of time. Expansion of partially saturated clays will continue after completion of the structure.

Evaluation

Most highway engineers strongly endorse the use of prewetting to minimize subgrade heaving. In view of the past experience and actual case studies, it is doubtful if prewetting can be successfully used with lightly loaded structures. The effective migration of moisture, the depth of penetration, the time required for saturation, and the swelling of partially saturated soils are not fully understood. Prewetting practice is much more complicated than assumed by most laymen. A great amount of research is required before complete evaluation of the prewetting practice can be made. Some of the disadvantages of the prewetting method are as follows:

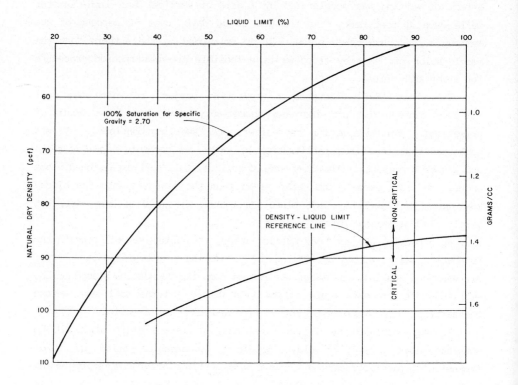

Figure 8-3. Clays encountered along San Luis Drain. (After Bara).

1. Allowing the ponding water to migrate into the lower moisture-deficient
    soils.   Experience indicates that in a covered area, the moisture
    content of the underslab soil seldom decreases. Wet soil will induce
    swelling.   After the swelling has reached its maximum potential,
    moisture migrates to the lower moisture-deficient soil and induces
    further swelling. This procedure can continue for as long as 10 years.

2. From a construction standpoint, the time required for prewetting can be
    critical.   A moisture condition of less than saturation is often
    adequate to inhibit objectionable uplift.   The length of ponding time
    required is usually about one to two months.   Even this length of time
    may be objectionable as being too great.

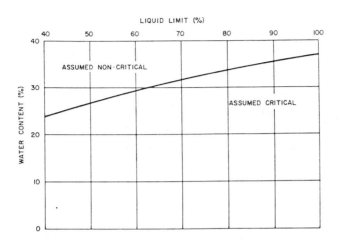

Figure 8-4. Minimum water content required for soil liquid limit.
(After Bara).

3. It is highly questionable if a uniform moisture content can be obtained
    in prewetted areas. Water can only seep into the stiff clay through
    fissures, and consequently, uniform distribution of moisture content is
    not likely to take place. As a result, differential heaving can be
    critical even after a prolonged period of prewetting.

4. Experiments indicate that ponding water can effectively penetrate the
    soil to a depth of four feet within a reasonable time. Such a depth is
    insufficient to provide a balanced moisture zone for the construction
    of important structures.

5. While prewetting may prove to be a possible method of stabilizing the
    soil beneath the floor slab, pavement, or canal lining, it is doubtful
    that footing foundations can be placed on prewetted soil. In saturated
    conditions, the bearing capacity of a stiff clay can be reduced to a
    very low value, less than 1,000 psf, which prohibits the use of
    conventional footing foundations.

While prewetting may play an important role in the construction of
slabs, it is doubtful that this method is an important construction
technique for building foundations on expansive soils.

COMPACTION CONTROL

The amount of swelling that occurs when a structural fill is exposed to additional moisture depends upon the following:
1.   The compacted dry density,
2.   The moisture content,
3.   The method of compaction, and
4.   The surcharge load.

The last two requirements are not critical in actual construction. The method of compaction is generally limited by the available equipment. For lightly loaded slabs, the surcharge load is usually very small.

Placement condition

As early as 1959, Dawson [8-7] suggested that highly expansive soils be compacted to some minimum density rather than to a maximum density.

Holtz and Gibbs [8-8] show the influence of density and moisture on the expansion of a compacted expansive clay, as shown on Figure 8-5.

It can be seen that expansive clays expand very little when compacted at low densities and high moisture but expand greatly when compacted at high densities and low moisture.

Gizienski and Lee [8-9] show that when their test soil was compacted at about 4 1/2 percent above optimum, which is 10 1/2 percent, the swell was negligible for any degree of compaction.

The main reason moisture content is important is that moisture content can generally result in low density fill, not that high moisture content will reduce swelling. The controlling element is density. Compacting stiff clay at four to five percent above optimum is very difficult. The process of recompacting swelling clays at moisture contents slightly above their natural percentage and at a low density is an excellent approach.

Referring to Chapter 2, it was established that the swelling pressure of clay is independent of the surcharge pressure, initial moisture content, degree of saturation, and thickness of stratum; it increases only with the increase of initial dry density. For instance, with reference to Figure 2-22 and Table 2-7, by decreasing the dry density of a typical expansive clay from 109 to 100 pcf, the swelling pressure decreases from 13,000 to 5,000 psf and the swelling potential decreases from 6.7 to 4.2 percent. All of this can be accomplished without changing the moisture content.

The main advantage of using this approach is that the swelling potential can be reduced without the adverse effects caused by introducing excessive moisture into the soil. Figure 2-16 indicates that to decrease the swelling potential from 6.7 to 4.2 percent, an increase of moisture content of about five percent will be required.

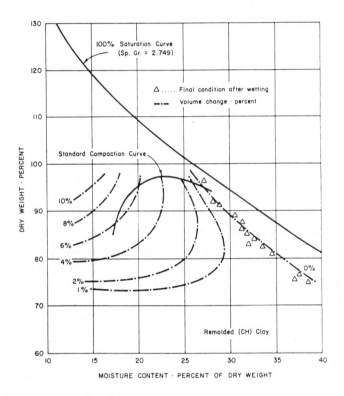

Figure 8-5. Percentage of expansion for various placement conditions
             when under unit psi load. (After Holtz and Gibbs).

The shortcomings of prewetting methods mentioned in the preceding section can be eliminated by compaction control. Excess water is not present in the soil; therefore, there is no migration of moisture to the underlying moisture-deficient soils; long waiting periods prior to

construction is unnecessary.   A reasonably good bearing capacity can be
assigned to the low density soil.

With modern construction techniques, it is possible to scarify,
pulverize, and recompact the natural soil effectively without substantially
increasing the construction costs.

Design

Leonard Kraynski of Woodward, Clyde & Associates suggests the following
design procedure on compaction control:

1.   An adequate mix should be prepared for three Proctor cylinders at each
     moisture content.   The cylinders are to be compacted using three
     different efforts, such as 12,400 ft.-lb. per cu. ft. (Standard
     AASHTO), 23,000 ft.-lb. per cu. ft., and 56,200 ft.-lb. per cu. ft.
     (Modified AASHTO).   Thus, a total of 12 to 15 samples will be adequate
     to define moisture-density curves as shown on Figure 8-6.

2.   From each compacted sample, a two-inch-diameter core may be extracted
     and tested in the consolidometer for swell.   The samples are subjected
     to 144-psf surcharge pressure, then submerged in water and allowed to
     swell.   From the measured percent expansion, curves of equal swell were
     plotted as shown on Figure 8-7.

3.   From a study of these results, a moisture content of 19 to 23 percent
     and a dry density ranging from 96 to 102 pcf were selected as design
     specifications.   Using the placement conditions, the average swell
     under a surcharge load of 144 psf is predicted to be five percent with
     a maximum swell potential of less than eight percent.   Such average and
     maximum swell are considered to be acceptable for the proposed type of
     construction.

4.   The required depth of compaction depends upon the degree of expansion
     and the magnitude of the imposed loads.   Generally, one to five feet of
     compacted material will be adequate with the range of two to three feet
     being the most commonly used.

Evaluation

The National Technical Service [8-10] summarized the effort of
mechanically altering the expansive soils to reduce their potential volume
change as follows:

"Ripping or scarifying the subgrade may be considered a minimal effort treatment alternative, since the depth of influence is not great (i.e., generally less than 2 ft) and the actual alteration of the soil is limited unless considerable effort is expended on mixing or remolding the soil. Ripping or scarifying is best suited for application to secondary highways since they can normally tolerate larger deformations and because the depth limitations of this treatment alternative preclude extensive alteration of the soil. However, this treatment may be considered for application to primary highways where the subgrade soils exhibit low potential swell and the uniformity of the final grade is of concern. If initial in situ moisture contents are low, then the ripping or scarifying should be followed by an application of water to increase the moisture content for volume change reduction purposes and to facilitate compaction. Compaction should be controlled so that the placement conditions minimize the eventual occurrence of volume change. Typical placement conditions for at least the upper six inches of the ripped or scarified layer should be 92-95 percent maximum dry density and optimum moisture content minus two percent or greater (AASHTO T-99)".

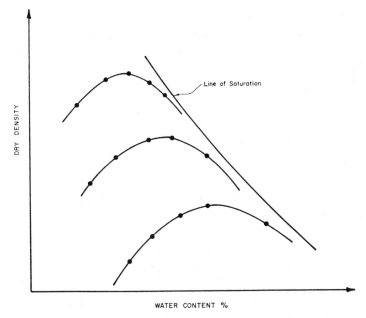

Figure 8-6. Preparation of specimens for earthwork specifications. (After Woodward-Clyde and Associates).

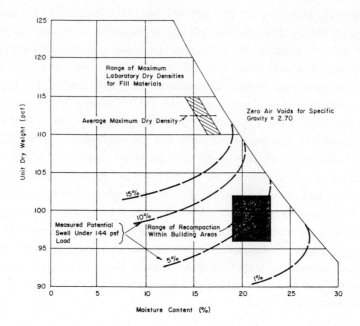

Figure 8-7. Determination of fill placement moisture and density.
(After Woodward-Clyde and Associates).

SOIL REPLACEMENT

A simple and easy solution for slabs and footings founded on expansive
soils is to replace the foundation soil with nonswelling soils. Experience
indicates that if the subsoil consists of more than about five feet of
granular soils (SC-SP) underlain by highly expansive soils, there is no
danger of foundation movement when the structure is placed on the granular
soils. The mechanics and the path of surface water seeping through the
upper granular soils and into the expansive soils are not clear. It is
concluded that either seepage water has never reached the expansive soils,
or the heaving of the lower expansive soils is so uniform that structural
movement is not noticeable.

This is not true in the case of man-made fill. For economic reasons,
the extent of the selected fill must be limited to a maximum of ten feet
beyond the building line.   Therefore, the possibility of edge wetting

exists. A guideline has not been established as to the thickness requirement for the selected fill. A minimum of three feet should always be insisted upon, although five feet is preferred. This thickness refers to the thickness of selected fill beneath the bottom of the footings or bottom of floor slabs.

The pertinent requirements concerning soil replacement are the type of replacement material, the depth of replacement, and the extent of replacement.

Type of material

Obviously, the first requirement for the replacement soil is that it be nonexpansive. All granular soils ranging from GW to SC in the Unified Soil Classification System may fulfill the nonexpansive soil requirement. However, for clean, granular soils such as GW and SP, surface water can travel freely through the soil and cause wetting of the lower swelling soils. In the other extreme, SC material with a high percentage of plastic clay sometimes exhibits swelling potential. The following criteria have been used with a certain degree of success:

Liquid Limit, percent	Percent minus No. 200 sieve
Greater than 50	15-30
From 30 to 50	10-40
Less than 30	5-50

It is becoming increasingly difficult to locate materials, fulfilling the above requirements, in expansive soil areas such as Metropolitan Denver. If necessary, the requirement for imperviousness can be forfeited. Any selected fill is satisfactory provided the material is nonexpansive. Also, swell tests are the only positive method of determining the expansiveness of the material. When in doubt, such tests should be conducted rather than relying on plasticity tests.

A great deal of emphasis has been given to the possibility of blending granular soil with the on-site swelling soils, thus reducing the amount of imported fill required. Theoretically, such a method is reasonable; but in practice it is difficult to incorporate granular soil with stiff, dry expansive clays. Disc harrows and plows are required to break the clay into reasonably sized clods. Such an undertaking is probably as expensive as using the lime stabilization method.

Depth of replacement

The depth of influence is a most complicated question that must be
answered when dealing with soil treatment beneath the slabs or footings. To
what depth should the natural soil be recompacted: How many feet of
overexcavation are required? How many cubic yards of nonexpansive soil have
to be imported? These questions cannot be intelligently answered until the
amount of movement that can occur beneath the slabs or footings can be
assessed.

Theoretically, the amount of uplift can be evaluated from the data
derived from swell tests and pressure distribution methods.   Gizienski and
Lee [8-11] evaluated the theoretically computed uplift derived from
laboratory test data and the actual measurement taken from a small scale
field test.   They found that the actual heave in the field was only one-
third of that estimated from the results of laboratory tests.

The Colorado Highway Department established curves which show the
relationship between total swell and the depth below the surface of the
subgrade [8-12].   Studies have shown that the swelling can take place down
to a depth of as much as 50 feet. Also, 60 percent of the swell in many of
the Colorado subgrade clays can occur down to a 20-foot depth.

While both the theoretical approach and actual measurement concerning
the depth of influence are urgently needed, the following should be pointed
out:

1.   The potential vertical rise of a soil mass, say 10x10x3-feet, (such as
     that used in Gizienski's experiment) under uniform saturation
     conditions, can be less than that of the same mass subject to local
     wetting only. Uniform wetting tends to equalize heaving.

2.   There is a definite gain in placing the structure on a nonexpansive
     soil cushion.   Even if the deep seated soils swell, the movement will
     be more uniform, and consequently, more tolerable.

3.   The depth of selected fill should never be less than 36 inches and
     preferably 48 inches. The swelling potential of the soil beneath the
     fill is very important because density and moisture conditions change
     at various locations.   It should be noted that with four feet of fill
     plus the weight of concrete, a uniform pressure of about 600 psf is
     applied to the surface of expansive soils.   For moderately swelling
     soil, such a surcharge load can be important in preventing potential
     heave.

4. The failure of the soil replacement method generally occurs during construction. If the subgrade or open excavation becomes wetted excessively before the placement of the fill, the trapped water causes heaving. In such a case, detrimental heaving occurs regardless of the thickness of the selected fill. The soils engineer should have the opportunity of supervising the placement of fill, or such a scheme should not be adopted.

5. The thickness of the imported fill can be reduced if a combination of the soil recompaction and soil replacement methods is used. The natural soil is scarified and recompacted as described under "Compaction Control" for a thickness of about two feet, then another two feet of selected compacted fill placed. The combined thickness of four feet should be adequate to control heaving.

6. The degree of compaction of the selected fill depends upon the type of supporting structure. For supporting slabs, 90 percent of standard Proctor density should be adequate. For supporting footings, a degree of compaction of 95 to 100 percent should be achieved.

Extent of replacement

The main reason that an artificially selected fill cushion is less effective than a natural granular soil blanket is that in natural conditions, the blanket extends over a large area, much larger than in the artificial condition. In an artificial fill situation, it is always possible for surface water to seep into the deep-seated expansive soil at the perimeter of the fill. Therefore, the larger the area of replacement, the more effective the fill.

Figure 8-8 shows the suggested extent of replacement for both basement and nonbasement conditions. With this arrangement, the possibility of surface water entering the foundation soil is greatly reduced. The type of material used for backfill should be the same as used for the underslab selected fill.

Evaluation

With present technology on expansive soils, soil replacement is the best method to use in obtaining a stabilized foundation soil. The following are the evaluations of the soil replacement method:

1. It is possible to compact the replaced nonexpansive soil to a high
    degree of compaction, thus enabling the material to support either
    heavily loaded slabs or footings. Such capability cannot be obtained
    by the prewetting method. Also, with the compaction control method, a
    high degree of compaction on expansive soils is not desirable, and,
    consequently, the load carrying capacity is limited.

2. The cost of soil replacement is relatively inexpensive when compared to
    chemically treating the soil. No special construction equipment, such
    as disc harrow, spreader, or mixer is required. The construction can
    be carried out without delay as is encountered in the prewetting
    method.

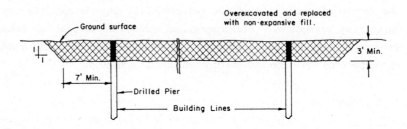

NON·BASEMENT CONDITION

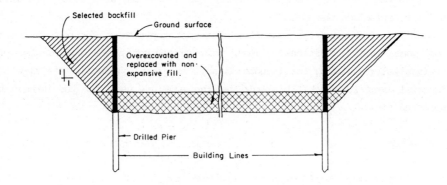

DEEP BASEMENT CONDITION

Figure 8-8. Suggested extent of fill replacement.

3. The granular soil cushion also serves as an effective barrier against the rise of ground water or perched water.

4. With the exception of a structural floor slab (suspended floor), soil replacement provides the safest approach to slab-on-ground construction.

5. To guard against unexpected conditions which might cause heaving, it is strongly suggested that floating slab construction be used. Slip joints must be provided for all slab-bearing partition walls so there is no chance of slab movement disturbing the structure.

6. Surface drainage around the building must be properly maintained so there is no opportunity for water to enter the expansive soils beneath the selected fill.

Current highway practice

The most obvious solution to highway engineers in resolving expansive soil problems is the removal of the undesirable soils and their replacement with non-expansive soils. The immediate concern is economic. How deep should the overexcavation be, how wide should the overexcavation be, and what is the availability of non-expansive soils? Considering the scope of the interstate highway, the above consideration becomes a multi-million dollar problem.

The Federal Highway Administration recommends the following guidelines [8-10]:

1. Minimum depth of application should be not less than two feet.

2. Material being put back should not be granular soils. Cohesive nonswelling soils such as silts, clayey silts, silty clays or some clays should be used.

3. The soils being removed can be used as backfill provided they are either chemically or mechanically altered to reduce the swelling potential.

4. Backfill should be compacted to 92-95% maximum dry density at optimum or greater moisture content (AASHTO T-99).

5. For interstate highways, the depths of subexcavation and replacement are:

Plasticity Index	Depth of Treatment, ft.
10-20	2
20-30	3
30-40	4
40-50	5
750	6

6. For secondary and state highways the guidelines are:

Plasticity Index	Depth of Treatment, ft.
10-30	2
30-50	3
750	4

7. Figure [8-9] show a typical example of the use of subexcavation and replacement for interstate highway.

The above guideline is practical and although the criteria is based on plasticity index only, it still provides an easy rule of thumb to follow.

J.R. Sallberg and P.C. Smith of the Highway Research Board [8-13] stated, "It is theoretically possible to estimate the depth below pavement surface which will require treatment or replacement in order to inhibit swelling." It was estimated for a certain site that 10 ft. of undercut would yield 40% reduction in swelling and 15 ft. of undercut would yield 50% reduction in swelling.

For a comfortable reduction of swelling on a high swell potential clay, the amount of removal on the order of six to ten feet is required. Much depends on the drainage and geometry of the pavement section.

A four-lane highway through expansive clay was proposed in Cyprus [8-14]. The Moni Formation is an outcrop only around the southern half of the island. The Moni clay is very "bentonitic" and can be as much as 90% montmorillonite-rich. The clay is green or greenish brown and highly fissured with abundant listric surfaces and relic shears. It varies from firm blocky soil to a brittle clay-shale, when first exposed. Moni clay classification properties are as follows:

Liquid Limit	77%
Plastic Limit	38%
In Situ Moisture	33%
In Situ Dry Density	122 pcf

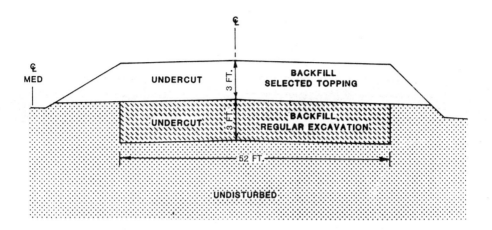

Figure 8-9. Typical example of subexcavation and replacement for an
             interstate highway.

Figure 8-10 shows the original and revised design sequence. The
procedure involves the use of an asphaltic membrane at the bottom of the
excavation, backfilled with nonexpansive soil and the installation of an
underdrain system.

LIME STABILIZATION

The use of lime to stabilize subgrade soil has been known to engineers
all over the world for a long time. For centuries, the Chinese have used
lime as a stabilizing agent in foundation soils.   Modern engineering
rejected the use of lime - in preference to cement - because the cementation
reaction of lime requires many months and the gain in strength is much
smaller than with cement. Since strength is not a requirement, lime is a
favorable agent to reduce the swelling potential of foundation soils. Most
of the lime stabilization projects were carried out by the highway
departments of various states.   For instance, the Texas State Highway
Department used nearly 1/2 million tons of lime for stabilization in 1969.
Although the success of lime-treated subgrade is questionable in many
instances, the use of the lime stabilization method has been steadily
increasing.

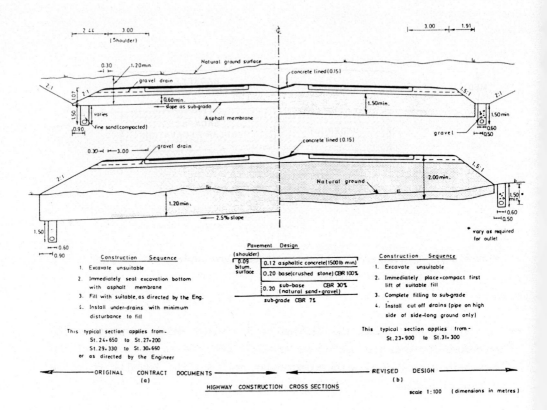

Figure 8-10. Highway construction cross-sections. (After Colliss).

Reaction

It is generally recognized that the addition of lime to expansive clays reduces the plasticity of the soil and, hence, its swelling potential. The chemical reaction occurring between lime and soil is quite complex. The stabilization apparently occurs as the result of two processes.

In one process, a base exchange occurs with the strong calcium ions of lime replacing the weaker ions such as sodium on the surface of the clay particle [8-15]. Also, additional non-exchanged calcium ions may be adsorbed so that the total ion density increases. The net result is a low base-exchange capacity for the particle with a resulting lower volume change potential.

In the other process, a change of soil texture through flocculation of the clay particles takes place when lime is mixed with clays. As the concentration of lime is increased, there is a reduction in clay content and a corresponding increase in the percentage of coarse particles. The reaction results in reduction of shrinkage and swell and improved workability.

W.G. Holtz [8-16] found that lime drastically reduces the plasticity index and drastically raises the shrinkage limit of montmorillonitic clays, as shown on Figure 8-11.

Application

The amount of lime required to stabilize the expansive soils ranges from two to eight percent by weight. The recently completed Dallas-Fort Worth Regional Airport [8-17] claims to have undertaken the world's largest lime stabilization project, consuming about 300,000 tons of lime. The subsoil consists of 8 to 16 feet of expansive clay with a potential vertical expansion equivalent to 10 percent of the layer thickness. The clays are underlain by shale of the Eagle Ford Formation.

The thickness of the treatment ranged from nine inches for taxiways and runways to 18 inches for aprons. For stabilization, six to seven percent of lime was required.

The stiff clay subgrade was broken down with a disc harrow to maximum sized clods of four to six inches. Lime was applied in slurry consisting of one part lime to two parts water by weight. The slurry was applied to the subgrade at 40 to 60 pounds pressure using water trucks. The application rate was sufficient to produce, within the stabilized layer, a dry lime content of six percent. Experience with this project indicated that the lime treatment not only transformed the soil to a nonswelling, friable mixture, but also improved the structural capacity of the treated layer.

In interstate highway construction in Florida, Oklahoma, and other states, lime stabilization was used to a large extent. In Oklahoma [8-18], a deep plowing technique was used. The subgrade was overexcavated two feet, then deep plowed by ripper-type equipment for an additional two feet. Lime was then added, and the deep plowing operation was continued until a good mix was obtained. After compaction, the two feet of soil which had been

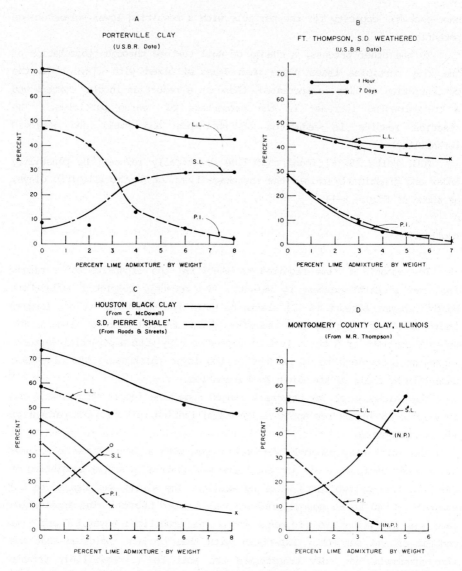

Figure 8-11. Effect of lime on plastic characteristics of montmorillonite
clays. (After Holtz).

removed was replaced in six-inch-thick layers, mixed with lime, and
compacted. The amount of lime used was about three percent by weight.

The successful use of mixing lime in expansive soils for highway and
airport construction is encouraging, although the depth of treatment

required and the results of the treatment on a long term basis has not been evaluated.

Recent studies indicate that sulfate-bearing clay soil can induce expansive reaction when mixed with lime. Lime treatment of Stewart Avenue in Las Vegas, Nevada, has induced heave in excess of 12 inches. Dal Hunter [8-19], engineering geochemist, stated, "With the present state of knowledge, lime-induced heave is difficult to predict for all but most obvious conditions."

Mixing lime in foundation soils to reduce swelling has not been seriously considered in the past. It appears that, with the knowledge gained from airport and highway construction using lime, treatment of underslab soils with lime deserves more attention. This is especially true in the case of large warehouses or school buildings where the floor covers a large area and a structural floor slab is not feasible due to the high cost. By overexcavating the site both in depth (three to four feet) and area and replacing the soil in compacted layers having adequate lime treatment, a stable slab can be expected. With present day limited knowledge of lime stabilization, footing foundations should not be placed on treated expansive soils.

In-place mixing

In the United States, various state highway departments have developed specifications for lime stabilized subgrade. These include the states of California, Louisiana, South Dakota, Illinois, Texas and Virginia. The basic lime stabilization process includes in-place mixing, plant mixing and pressure injection.

In-place mixing can be accomplished by mixing lime with existing material excavated from the subgrade. Lime can also be incorporated into the borrow material either at the borrow site or at the construction site. The amount of lime required can be estimated by various methods. Eades & Grim [8-20] suggested the lime pH test can provide an estimate of the percent lime required to effectively reduce the plasticity of a soil. AASHTO methods T-89 and T-90 are utilized to determine the liquid limit, the plastic limit and the plasticity index of the soil treated with various percentages of lime. The design lime content may be designated as that lime content above which no further appreciable reduction in PI occurs or a minimum lime content which produces an acceptable PI reduction. The Louisiana and Texas procedures are based on the increase of unconfined compressive strength as specified by AASHTO T-220. A more complicated and detailed analysis is given by Thompson [8-21]. The Thompson procedure has been published by ASTM special technical publication.

Design lime contents generally are based on an analysis of the effect of varying lime percentages on selected engineering properties of the soil-lime mixture. Different design lime content for the same soil may be established depending on the objectives of the lime treatment and mixture design procedure utilized.

The following procedures are utilized for in-place mixing:

1. One increment of lime is added and the mixture is allowed to mellow for a period of one to seven days to assist breaking down of the heavy clay soils.

2. Two increments of lime are added for soils which are extremely difficult to pulverize. Between the applications of the first and second increments of lime, the mixture is allowed to mellow.

3. For deep stabilization, one increment of lime is applied to modify soil to a depth of 24 inches. A second increment of lime is added to the top one to twelve inches for complete stabilization. Plows and rippers are used to break down the large clay chunks in the deep treatment.

Pressure injection

The pressure injection method of lime stabilization has been used in Jackson, Mississippi, in Calexico, California, and in Tucson, Arizona. The method consists of pressure injecting lime-water slurry into the soil through closely spaced drill holes as shown on Figure 8-12.

The drilled holes were five feet deep, located adjacent to the building, and on three-foot centers. In Jackson, Mississippi, where the soils beneath 200 houses were treated, it was reported that an estimated 10 percent of the treated soils had to be retreated, and one percent received three treatments.

The movement of slurry through fine-grained or clay soils is through the cracks, fissures or other discontinuities. Actual diffusion of the lime into the soil is very limited. Most of the calcium reaction occurs at the surface of the cracks or fissures that the lime slurry flows through.

Proponents of the pressure injection method claimed that lime slurry pressure injection can create a stabilized moisture barrier and become an impediment to the movement of moisture through the treated area of clay. Specially designed equipment is used to pump the slurry into the voids, fissures, seams and cracks of the clay from 1 1/2 to 8 feet below the surface.

This method of treatment has been used as a remedial process, where damaging movements of the soils have occurred and also as a preconstruction process to stabilize soils of known expansive qualities for residences, commercial buildings and parking areas.

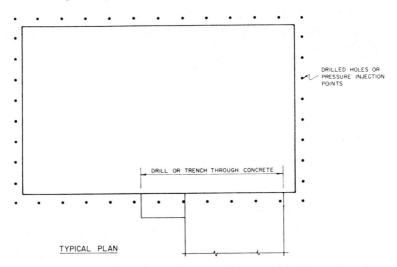

DRILLED HOLES OR
PRESSURE INJECTION
POINTS

DRILL OR TRENCH THROUGH CONCRETE

TYPICAL PLAN

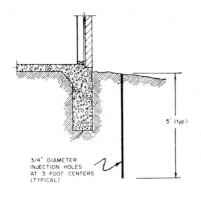

5' (typ.)

3/4" DIAMETER
INJECTION HOLES
AT 3 FOOT CENTERS
(TYPICAL)

1) INJECT SLURRY USING TWO PIPE SYSTEM -
    OUTER PIPE 3/4 INCH DIAMETER, POINTED AT
    BOTTOM AND PERFORATED IN LOWER FOOT
    WITH 1/8" HOLES; INNER PIPE IS 1/4" IN.
    DIAMETER. THE PIPES ARE JETTED IN THE
    GROUND.

2) CONTINUE TO INJECT SLURRY IN EACH HOLE
    UNTIL SLURRY COMES OUT OF GROUND AROUND
    THE PIPE. REPORTED INJECTION PRESSURES AT
    NOZZLE ARE IN THE RANGE OF 200 TO 400 PSI

3) TYPICAL SLURRY PROPORTIONS: 50 SACKS OF
    HYDRATED LIME (50 LBS./SACK) TO 900
    GALLONS OF WATER; $Ca(OH)_2$ CONTENT IN
    LIME AVERAGES 95%. LIME AND WATER SLURRY
    ARE MIXED IN A BLENDING TANK PRIOR TO
    INJECTION.

DETAIL

Figure 8-12. Lime stabilization - pressure injection method.
(Calexico, Calif. & Jackson, Miss.)

It is difficult to deny that pressure injection has certain potential applications in the expansive soil problem area, but by no means is it a cure-all for the expansive soil problem [8-10].   Before such a method is used, it should be determined whether the soil is reactive with lime and whether the lime slurry can be injected in the soil.

L.K. Davidson [8-22] stated in 1965 that the results of laboratory studies show that lime does diffuse into a soil-water system.   For the experimental conditions, the rate of diffusion was very slow and given by the equation:

$$L = 0.081 \ t^{1/2}$$

Where: L = lime penetration distance, in.

t = time, days

Using this formula results in a penetration distance of 1.5 inches in one year.

It is the conclusion from both laboratory and field experience that lime migration into expansive soils is extremely slow.   The rate of migration can probably be increased by introducing large quantities of water to carry the lime slurry.   There is the potential danger of triggering an excessive amount of swelling in the deep seated soils.

Woodward-Clyde-Sherard & Associates [8-23], in their investigation, concluded that the success of lime treatment is probably because of moisture barrier effects rather than because of any widespread changing of soil properties.

Street pavement

The use of lime for subgrade treatment in highly expansive soil areas in south Denver has experienced limited success.   At Harriman Park in south Denver, the problem of swelling soil has attacked more than one-third of the residences, Figure 8-13.   Overall, the distress in the roadways could be generalized as numerous vertical displacement in the roadways ranging from several inches to two feet (Figure 8-14).   Transverse cracks were noted with separations up to one inch.   The most significant feature of the roadway distress is the rolling nature of the road.   Local residents described the effect as a "roller-coaster" ride.   The cause of distress was generally blamed on leakage of sewerlines, but it was not possible to determine whether the leakage was caused by differential heaving or the heaving was caused by leakage.

The depth of bedrock is generally shallow, ranging from zero to ten feet below grade. In most cases, the basements of the houses are placed on bedrock. Bedrock consists of Pierre Shale (Upper Cretaceous) with sandstone lenses. The sandstone lenses were observed to stand at 85° to 90° from

Figure 8-13. Sidewalk heaving indicating the magnitude of swelling pressure exerted along this section of pavement.

horizontal. The claystone portion of bedrock possesses high swell potential, as much as six percent under a surcharge load of 500 psf and swelling pressure in excess of ten ksf.

The change of pH value with the increase of the lime content was plotted, which yielded a curve from which the optimum percentage of lime that will react with the soil was selected, in this case, four percent as shown in Figure 8-15. It is interesting to note that with the addition of four percent of lime in the soil, the swelling potential decreased substantially, but the swelling pressure remains on the order of ten ksf.

Actual reconstruction of the pavement in the subdivision consists of a combination of subgrade modification and lime treatment as follows:

1. Overexcavating and reworking of the upper three feet of subgrade material.

2. The on-site material was mixed and replaced in maximum eight-inch lifts compacted to 90% to 92% of standard Proctor density at last two percent over optimum moisture.

Figure 8-14. Pavement deterioration due to heaving.

3. A minimum eight-inch layer of lime stabilized subbase with a lime content of four percent by weight should be placed beneath the pavement section. The stabilized subbase should extend the full width of the pavement section and two feet beyond the curbs. The subbase should be placed at a moisture content of optimum to three percent above optimum and compacted to 95% standard Proctor density. This section will

provide a semi-impervious layer for reducing subgrade moisture variations and also provide uniformity of the subgrade.

The reconstructed lime stabilized pavement has been in service for more than three years. There were no complaints from the residents. The pavement was in fair condition, certainly much better than the condition experienced prior to reconstruction. The wavy condition still existed in some areas, indicating there still is subgrade heaving.

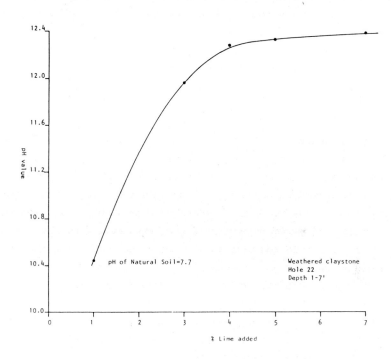

Figure 8-15. pH value and lime content.

CHEMICAL STABILIZATION

Besides the use of lime, other chemicals, both organic and inorganic, can be used to stabilize expansive soils. Cement and fly ash have both been used in the laboratory with successful results. Of course, the cost of

cement is considerably more than that of lime. Fly ash is sometimes added
to the soil-lime mixture to increase pozzolanic reaction.

Ion exchange by the addition of divalent salts, cation fixation in
expanding lettice clays with potassium, deactivation of sulfates with
calcium chloride, waterproofing with silicones or asphalts, cementation with
silicates, carbonates, lignins, phosphoric acid and alteration of
permeability and wetting properties with surface active agents have all been
used to reduce the expansive characteristic. All the above chemicals have
proved to have total or limited success in the laboratory, essentially by
reducing the plasticity index. Some chemicals can virtually alter the soil
from plastic to non-plastic [8-24].

Most of these chemicals are effective under laboratory conditions, but
their application in the field is very difficult. There is no supporting
evidence that any of the chemicals has economically worthwhile benefits
[8-25].

Cement stabilization

The hydration products of portland cement include calcium silicate
hydrates, calcium aluminate hydrates, and hydrated lime. During hydration,
portland cement releases a large amount of lime. It is believed that the
base exchange and cementing action of portland cement with clay are similar
to that of lime. In addition to the above actions, the incorporation of
portland cement in clay increases the strength of the mixture. The
resulting product commonly known as soil-cement is familiar to most soil
engineers.

The action of cement on clay minerals is to reduce the liquid limit,
the plasticity index, and potential volume change, and to increase the
shrinkage limit and shear strength [8-26].

Spangler and Patel [8-27] reported on the laboratory treatment of an
expansive Iowa gumbotil with portland cement. The addition of two percent
and four percent of portland cement considerably reduced the potential
volume change of the soil.

Jones [8-15] added two to six percent of portland cement to the
expansive Porterville clay of California which resulted in the pronounced
reduction of volume change characteristics.

The effect of cement and of lime was about the same in reducing soil expansion, but the cement reduced the shrinkage of air-dried specimens about 25 to 50 percent more than did the lime.

The mixing and dispersing methods for cement are nearly identical to those for lime. The difficulties of uniformly introducing portland cement into very fine-grained soils are generally greater than with lime because it is less soluble.

Both cement and lime have been used in highway construction for modifying the swelling property of the subgrade soil. The use of cement and lime to stabilize underslab soil in buildings is seldom reported. There appears to be a great potential for using cement to modify the underslab soils. With two to six percent incorporated in the clay, the resulting soil-cement mixture acts as a semi-rigid slab. If the deep-seated soil expands, the swelling effect tends to distribute uniformly, this reducing damage caused by differential heaving. Such construction is particularly favorable for the treatment of a large warehouse floor where a crack-free, level floor is essential and the use of structural floor slab is economically prohibitive. Due to lack of strength, the use of lime cannot provide a semi-rigid element beneath the slab.

A great deal of research and field study is required before cement stabilization can be economically applied. An effective application method, either by mixing or by slurry injection, must be perfected before the scheme can be considered in practice.

Organic compound

Organic compounds stabilize expansive soils by waterproofing, by retarding water absorption, or by hardening the soil with resins. Organic compounds such as Arguard 2HT or 4-Terf-Butylpyrocatechol have been used with a limited degree of success.

Davidson and Glab [8-28], in laboratory investigation of highly plastic Iowa subgrade soils, have shown that certain organic compounds which furnish large organic cations when dissolved in water have considerable promise as admixtures to increase the stability of such soils. They found that water solutions of chemical admixtures of this type decreased plasticity, shrinkage, and swelling of plastic soil samples.

A proprietary liquid known as Fluid 705, 706, and 707 was introduced by Soil Technology Corporation in Denver, Colorado. The fluid was mixed with swelling clays and tested in the laboratory for physical characteristics, swelling potential, and permeability.

Expansion tests were performed on three remolded specimens of Denver clay shale. One specimen was treated with distilled water, the second with proprietary fluid 705, and the third with proprietary fluid 706. A surcharge load of 100 psf was applied to each specimen. The specimens were saturated with distilled water and the amount of expansion determined. The specimen treated with fluid 705 did not expand. The specimen treated with fluid 706 was moderately expansive. The specimen treated with distilled water was highly expansive. In both cases, the addition of proprietary fluid has changed its Atterberg limits from high plasticity to nonplastic.

The permeability tests were performed on specimens comprised of a mixture of 15 percent clay and 85 percent silica sand by weight. The sand used was Silica Sand Natural Grain, supplied by the Ottawa Silica Company, Ottawa, Illinois. The clay used was Aquagel supplied by the Baroid Division of the national Lead Company, Houston, Texas.

Constant head permeability tests were performed on two remolded specimens of clay and silica sand using distilled water in one test and the proprietary Fluid 707 in the other. The material was found to be impervious to distilled water during the 34-day testing period. The coefficient of permeability of the proprietary Fluid 707 treated soil was determined to be approximately six feet per year.

Permeability test is important, as in actual application the fluid must be able to migrate into the soil. The ability of the fluid to permeate in the impervious soil is encouraging.

The first large-scale experiment on the use of the proprietary fluid took place in December 1974 in Denver, Colorado. Specially designed equipment as shown on Figure 8-16 was used. The machine could hydraulically bore 1-1/2-inch-diameter holes, three at a time, to a depth of more than 10 feet in stiff clay and claystone shale. The auger was advanced by a pressure of 300 to 500 psi. Thus, the holes could be advanced a total of 10 feet in less than a half minute. The holes were spaced 36 inches apart and in highly impervious soil the spacing was reduced to 18 inches. Proprietary fluid was introduced into the holes under a pressure of about 10 psi.

Figure 8-16. Equipment used for chemical injection.

The treatment was intended to extend for a depth of at least
six feet.   It was intended to reduce the plasticity index of the expansive
clays from about 40 to 10 percent and the swelling potential from moderate
swelling to nonswelling.   Undisturbed soil samples were taken before and
after treatment to determine the effectiveness of the application.   The
results were not as expected.   Both the plasticity index and the swell
potential did not significantly reduce.   Valuable experience was gained from
the experiment, however, some of which follows:

1.   The holes should have a maximum spacing of 12 inches.
2.   The fluid must be applied under a pressure of not less than 250 psi.
3.   Pressure gauges should be provided to indicate a pressure drop when the
     fluid flows into the seams and fissures in the clay. The auger should
     then be advanced to avoid the fissures.

It is believed that with further study on field application and
mechanical improvement, the above method will eventually find an important
place in the realm of chemical stabilization.

Heat treatment

Studies carried out in India [8-29] have shown that the plasticity
index of black cotton soils decreases as temperature increases until at 500°
C, soil becomes nonplastic. Field experiments were undertaken with a mobile
furnace. It was found that the effective depth of burning of the machine is
hardly 2.5 inches which reduces the output and, consequently, increases the
cost of production. It was observed that the technique was not very
economical due mostly to the high prevailing cost of fuel.

Explosive treatment

Explosive treatment is an experimental method used by the Colorado
State Highway Department [8-30] to replace subexcavation and recompaction.
Explosive treatment can reduce the density of expansive shale,
disorientation of bedding and a substantial addition of water, thus
decreasing the swelling potential.

The area treated is located on I-70 north of Grand Junction,
Colorado. In this area, a 1,150-foot cut section contained active swelling
Mancos Shale, which has required extensive periodic maintenance since the
road was completed in 1965. The blast holes were eight feet deep, spaced at
seven-foot intervals and loaded with 1/3 pound of 90% dynamite and three
pounds of NNFO. The blasting has significantly lowered the density of the
shale from 130 to 150 pcf dry density before blasting to 90-110 pcf
following treatment. Blasting has caused a uniform rise in the treated area
of three to four inches.

The cost of treating this 1,150-foot cut using a low-level explosive
was approximately one-fifth the estimated cost of subexcavation and
recompaction. Initial evaluation indicated that the treatment has performed
very well. Since blasting would cause cavities in the subgrades, settlement
and distortion of the roadway could take place. Long-term observation is
required to establish the effectiveness of this treatment.

MISCELLANEOUS STRUCTURES

All structures founded on expansive soils are subject to heaving movement. Almost all research and investigations were directed to building foundations and highway pavement. It is the miscellaneous structure founded on expansive soils that offers the most challenge to the geotechnical engineer. Such structures include retaining walls, buried conduits, canal linings, swimming pools, water treatment reservoirs, transmission towers and others. Very little has been written on such structure foundations, and yet failures are frequently reported. As an example, micro-wave towers are sensitive to movement. Heaving of foundation soil can throw the beam off its target and cause transmission problems over a large area.

Some of the more commonly encountered structures are discussed as follows:

Buried pipes

As early as 1962, Kassiff and Zeitlen [8-31] found that high stresses resulted in the pipe through inequalities in the lateral and vertical swelling behavior of the clay, and that these stresses could be greater than those caused by internal pressure. Rapture of cast iron, concrete and asbestos pipes buried in swelling clays were frequently reported.

The Del City Pipeline near Oklahoma City [8-32] is a 6.2-mile long rubber gasketed concrete pipe, varying in diameter from 20 to 24 inches. From July, 1965 to June, 1979, 116 leaks were detected. The subsoils at places consist of stiff clays with moderate-to-high swell potential. The following evaluations were made by the researchers:

1. In the expansive soil area, it is desirable to use flexible pipes instead of rigid pipes. Flexible pipes can sustain relatively large deflection on the order of three percent to five percent without failure.

2. When rigid pipe is used, the strength of the pipe in bending should be higher than that required for construction under usual, nonexpansive soil conditions.

3. It is desirable to overexcavate a portion of the expansive soil and replace it with nonexpansive material, so that a more uniform heaving can take place.

4.   The pipe and joints can be designed to be flexible enough to move with
the expansion of the foundation without failure.

The water conveyance pipeline usually started with only a small leak.
Subsequently, water seeped into the expansive soil and caused large
differential heaving and large damage occurred. It is, therefore, important
that early detection and repair be made.

At Ankara, Turkey, [8-33] a new water line consisting of precast
reinforced concrete pipe, 60 inches in diameter, heaved about four feet in a
uniformly increasing manner from the restrained end to the free end of
construction along a segment of about 130 feet. The heaving occurred within
a short period upon the diversion of a small stream crossing the pipeline
route. The soil profile was of "Ankara clay" with a liquid limit of 60%.

The heaved pipe section was removed and the new pipe was imbedded in
gravel and about six feet of compacted fill was placed on top of it. The
idea was to prevent surface water from entering into the subsoil and to
carry away through the gravel any infiltration water.

Canal Lining

One of the most difficult structures to design against expansive soil
is the canal lining. Unlike swimming pools or water storage reservoirs
which occupy only a small area and where extensive treatment can be imposed
on them to prevent swelling, a canal usually runs for a long distance and
elaborate treatment generally is not economically feasible.

The most well-documented project in the United States is the
Friant-Kern Canal in California [8-34]. Approximately one-third of the
length of the canal, about 54 miles, transverses an area of expansive
clay. The canal began experiencing cracking, sliding and sloughing of the
side slope in both the concrete and earth-lined section. In 1970, the
Bureau of Reclamation decided to remove a portion of canal lining, flatten
the canal slopes and reline the canal using a compacted soil-lime mixture in
an attempt to stabilize the slopes.

The clay has a liquid limit ranging from 57 to 70, a plasticity index
from 37 to 46 and a shrinkage limit approximately of eight. For
stabilization of the expansive clay soil, four percent granular quicklime
was added to the soil. The addition of lime reduced the plasticity index of

the soil to about ten, with a significant increase in compacted soil-lime strength and improving the workability of the material.

Since the rehabilitation, the Friant-Kern canal lining has not experienced any failures in the areas where compacted soil-lime was used. The soil-lime lining is extremely erosion resistant. The erosion was, in general, restricted to less than 0.1 feet after up to six years of operation. Figure 8-17 shows the typical cross-sections.

A 140 kilometer portion of the right bank canal of the Malaprabha project in India passes through black cotton soil deposits [8-35]. The depth of cutting of the canal varies from one to ten meters, and the height of the bank varies from zero to a maximum of fifteen meters. Even before water was allowed for irrigation, it was found that the banks of the canal had failed and the lining had collapsed. Figures 8-18 and 8-19 show the collapsed banks.

A cohesive non-swelling soil layer was placed on the bottom of the canal immediately after excavation. On the 2:1 side slopes a horizontal component of cohesive non-swelling layer one meter thick was placed, wider than geometrically required. The provision of this layer of material is helpful in retaining shear strength in expansive soils after saturation.

A similar problem was experienced in Pingtin Shan, Nanyang China, where the North-South water transfer project has taken place. Engineers are still studying the most economical method in dealing with the canal lining in expansive soils.

Swimming Pool

The design of a swimming pool in an expansive soil area offers the most difficult challenge to a geotechnical engineer. Both indoor and outdoor pools are constructed in the residential area by pool contractors without considering the subsoil swelling problem. All pool manufacturers including those of gunite, figerglass, and aluminum claim the pool does not leak. In fact, all pools leak with time in varying amounts. A small amount of leakage can cause the subsoil to swell and result in pool or slab movement, which widens existing cracks and introduces more water into the subsoils, thereby compounding the problem.

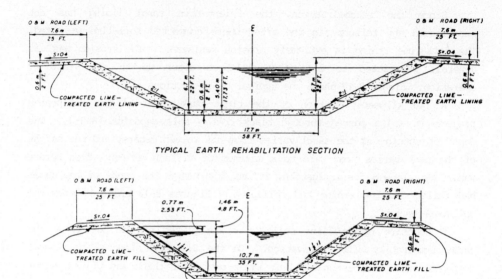

Figure 8-17.   Typical concrete-lined rehabilitation section.

Unlike large reservoirs, highway subgrade or airport runways, large-scale soil modification operations cannot be easily accomplished. The author's experience indicates that water leaking from the pool or from the apron through the backfill must be carried out immediately before extensive damage is done. The following construction sequences are suggested:

1.   The pool should be designed and constructed to withstand minor differential movement without serious cracking.

2.   Natural soil below the pool should be overexcavated to a depth of at least four feet and replaced with nonexpansive fill material compacted to 95% of the maximum standard Proctor density at about two percent above optimum moisture content.

Figure 8-18.   Heaving of canal bottom

3. An impervious membrane should be provided on the walls of the
    excavation and on top of the compacted fill to help prevent moisture
    from migrating into the expansive soils.

4. Asphalt membranes constructed from catalytically blown asphalts (see
    Table 7-1) can be effectively used.   The use of other forms of
    rubberized asphalt or PVC can also be considered.

5. A minimum eight-inch free-draining gravel should be placed beneath the
    pool slab, and below the pool deck.   This drainage layer should slope
    to a drain line or collection point from which water can be removed by
    pumping.   The drainage layer under the deck should slope to a perimeter
    drain or be connected to the under-pool layer by free-draining
    backfill.

6. The drains should consist of perforated pipe surrounded by a minimum of
    12 inches of free-draining granular material.

7. A light joint should be provided between the pool and deck so water
    splashed from the pool will not infiltrate the subsoils.

Figure 8-19. Shifting of concrete lining due to expansion
of black cotton soils

A chemical reaction may cause deterioration of pool slab and wall, which is unrelated to expansive soil, but will introduce large leakage. A major aluminum pool was totally destroyed by corrosion caused by cathodic action.

REFERENCES

[8-1]   McDowell, C., "Remedial Procedures Used in the Reduction of Detrimental Effect of Swelling Soils," Texas Highway Department.

[8-2]   Stevens, J.B., and Matlock, H., "Observation of Expansive Clays in Roadways," Transportation Research Board, 1986.

[8-3]   "Recommended Practices for Construction of Residential Concrete Floors on Expansive Soil," Portland Cement Association Vol. II, Los Angeles, California, 1970.

[8-4]   Bara, J.P., "Controlling the Expansion of Desiccated Clays During Construction," Second International Research Conference on Expansive Clay Soils, August, 1969.

[8-5]   Blight, G.E., and Wet, J.A., "Acceleration of Heave of Structures on Expansive Clay," Moisture Equilibria and Moisture Changes in the Soils Beneath Covered Areas.

[8-6]   Felt, E.J., "Influence of Vegetation on Soil Moisture Content and Resulting Soil Volume Changes," Proceedings, Third International Conference on Soil Mechanics and Foundation Engineering, Zurich, Vol. I, 1953.

[8-7]   Dawson, R.F., "Modern Practices Used in the Design of Foundations for Structures on Expansive Soils," Quarterly, Colorado School of Mines, Vol. 54, No. 4, 1959.

[8-8]   Holtz, W.G., and Gibbs, H.J., "Expansive Clay-Properties and Problems," Quarterly of Colorado School of Mines, Vol. 54, No. 4, October, 1959.

[8-9]   Gizineski, S.F. and Lee, L.J., "Comparison of Laboratory Swell Tests to Small Field Tests," Concluding Proceedings, International Research and Engineering Conference on Expansive Clay Soils, Texas A and M Press.

[8-10]   U.S. Department of Commerce, National Technical Information Service. PB80-139660 "Technical Guidelines for Expansive Soils in Highway Subgrades," 1979.

[8-11]   Gizienski, S.F. and Lee, L.J., "Comparison of Laboratory Swell Tests to Small Scale Field Tests," International Research and Engineering Conference on Expansive Soils, 1965.

[8-12] "Lime Shaft and Lime Tilled Stabilization of Subgrades in Colorado Highways," Interim Report 1967, Planning and Research Division, Dept. of Highways, State of Colorado.

[8-13] Sallberg, J.R., and Smith, P.C., "Pavement Design over Expansive Clays: Current Practice and Research in the United States," Highway Research Board, 1965.

[8-14] G.I. Colliss, "Preliminary Observations on the Design and Construction of a Four-Lane Highway Through Expansive Clay in Cyprus," Quarterly Journal of Engineering Geology, London, Vol. 17, 1984.

[8-15] Jones, C.W., "Stabilization of Expansive Clay with Hydrated Lime and Portland Cement," Bulletin, Highway Research Board, No. 193, 1958.

[8-16] Holtz, W.G., "Volume Change in Expansive Clay Soils and Control by Lime Stabilization of Expansive Clays at the Dallas-Fort Worth Airport," Proceedings of Workshop on Expansive Clays and Shale in Highway Design and Construction.

[8-17] Kelly, J.D., "Lime Stabilization of Expansive Clays at Dallas-Fort Worth Airport," Proceedings of Workshop on Expansive Clays and Shale in Highway Design and Construction, Vol. 2, 1973.

[8-18] Thompson, M.R., "Lime Stabilization: Deep Flow Style," Road and Streets, March, 1969.

[8-19] Hunter, Dal, "Lime-Induced Heave in Sulfate-Bearing Clay Soils," Journal of Geotechnical Engineering, Vol. 114, No. 2, ASCE 1988.

[8-20] Eades, J.L., and Grim, R.E., "A Quick Test to Determine Lime Requirements for Lime Stabilization," Highway Research Record 139, Highway Research Board, 1966.

[8-21] M.R. Thompson, "Suggested Method for Mixture Design Procedure for Lime-Treated Soil," Special Procedures for Testing Soil and Rock for Engineering Purposes, A.S.T.M. STP 479, 1970.

[8-22] Davidson, L.K., Demirel, T., and Hardy, R.L., "Soil Pulverization and Lime Migration in Soil-Lime Stabilization," Highway Research Board, 1965.

[8-23] "Remedial Methods Applied to Houses Damaged by High Volume-Change Soils," Woodward-Clyde-Sherard & Associates, FHA Contract H-799.

[8-24] Mitchell, J.K., and Raad, L., "Control of volume Changes in Expansion Earth Materials," Proceeding, Workshop on Expansive Clays and Shales in Highway Design and Construction Vol. II, 1973.

[8-25] Gramko, G.J., "Review of Expansive Soils, " Journal of the Geotechnical Engineering Division, June, 1974.

[8-26] Croft, J.B., "The Influence of Soil Mineralogical Composition on Cement Stabilization," Geotechnique, London, England, Vol. 17, June, 1967.

[8-27] Spangler, M.G., and Patel, O.H., "Modification of a Gumbotil Soil by Lime and Portland Cement Admixtures," Proceedings, Highway Research Board, Vol. 29, 1949.

[8-28] Davidson, D.T. and Glab, J.E., "An Organic Compound as a Stabilization Agent for Two Soil Aggregate Mixtures," Proceedings Highway Research Board, Vol. 29, 1949.

[8-29] Uppal, H.L., "Modification of Expansive Soils for Use in Road Work," Transportation Research Board, 1986.

[8-30] LaForce, Robert, "Explosive Treatment of Correct Swelling Shale," Colorado Department of Highways Report No. CDH-DTP-R-81-8, 1981.

[8-31] Kassiff and Zeitlen, "Behavior of Pipes Buried in Expansive Clays," Journal of Soil Mechanics and Foundation Division, ASCE, Vol. 88, Paper No. 3103, April, 1962.

[8-32] Rocklin, Robert C., "Water Conveyance Pipelines in Expansive Soil," 4th ICOES, 1980.

[8-33] Ordemir, Soydemir and Birand, "Swelling Problem of Ankara Clay," Proceedings of the 8th International Conference on Soil Mechanics and Foundation Engineering.

[8-34] Byer, Jack G., " Treatment of Expansive Clay Canal Lining," 4th International Conference on Expansive Soils, Denver, 1980.

[8-35] "Construction of Malaprabha Right Bank Canal in Black Cotton Soil Area," Government of Karnatak, Irrigation Department, India, 1987.

Chapter 9

SOIL SUCTION

INTRODUCTION

The concept of soil suction was initiated by agriculture scientists and has been used by soil engineers for many years. As early as 1956, J.W. Hilf [9-1], in his study of pore water, proposed to use triaxial test techniques to measure the suction of soil samples taken from the field. W.R. Wray [9-2] stated, "Soil suction is being increasingly used to explain expansive soil behavior, but its concept is not often addressed in geotechnical engineering textbooks." The soil suction concept has been widely advocated in Australia, Canada, South Africa and Israel. In the United States, extensive research has been conducted by L.D. Johnson [9-3], D. Snethen [9-4], J. Nelson [9-5] and others. Although the actual application of soil suction in geotechnical consulting practice is still limited, the term "soil suction" is used more and more often in expansive soil studies.

WHAT IS SOIL SUCTION?

Soil suction can be described in laymen's terms as a measure of a soil's affinity for water. In general, the drier the soil, the greater is the soil suction [9-2]. Soil suction is a parameter describing the state of the soil and indicates the intensity with which it will attract water. For practical purposes, the suction of a soil is considered to consist of two parts, osmotic suction and the matric suction. The sum of these two parts is termed the total suction. Thus, the total suction, h, is given by

$$h = h_o + h_c$$

where h_o is the osmotic suction

h_c is the matric suction

Osmotic suction

Osmotic suction in a clay is related to forces from the osmotic repulsion mechanism arising from the presence of soluble salts in the soil water and is identical in context with the osmotic attractive forces

[9-6]. To discuss the osmotic suction, it is convenient to consider a pure water placed in contact with a solution through a semi-permeable membrane that allows the movement of water molecules but does not allow the passage of solute. Because of the concentration of the solution, there is a tendency for the flow of water into the solution through the semi-permeable membrane. If flow is restricted, a pressure differential between the solution and pore water results. This pressure differential is the osmotic pressure, π, as given by

$$\pi = \Omega RTC_s$$

where Ω = the molal osmotic coefficient of the solute

R = the universal gas constant

T = the absolute temperature

C_s = the concentration of the solute.

It is well recognized that osmotic pressure can be expected to take place in the soil-water system. Assuming that the double layer system exists in the soil lattice, the concentration of ions being held by the attractive force prevents the ions from moving away from the double layer. However, water is able to move in and dilute the concentration, and, consequently, a semi-permeable membrane effect is achieved, as shown on Figure 9-1.

Matric suction

Figure 9-2 shows an idealized air-water interface in the connecting channel between two pore spaces in a soil. This idealized surface is similar to the water surface in a capillary tube. By equating the force due to the interfacial surface tension with that due to the difference between the air and water pressures, it can be shown that [9-7]

$$h_c = (u_a - u_w) = \frac{2T_s}{r}$$

where T_s = the surface tension

r = as shown in Figure 9-2

u_a = pore-air pressure

u_a = pore-water pressure

In addition to the surface tension forces, there also exist adsorptive forces exerted on the water molecules by the surface of the soil particles. These adsorptive forces account for the fact that the curvature of the water film is actually in the opposite direction adjacent to the individual particles as shown in Figure 9-3. The adsorptive forces allow for relatively high tensile stresses to be generated in the soil water that can be significantly greater than one atmosphere.

The component of the suction described by the above equation is termed the matric suction, h_c.

Although Figure 9-2 represents an idealized situation, it is evident that if the difference between air and water pressure were increased, the radius of the interface must decrease. Thus, as the matric suction increases, the air-water interface must recede to smaller pore spaces. This, of course, necessitates drainage of water from the soil. Consequently, there is a unique relationship between the water content of the soil and the matric suction. Obviously, the relationship will depend upon the distribution of pores and their sizes and shapes, i.e., the soil fabric.

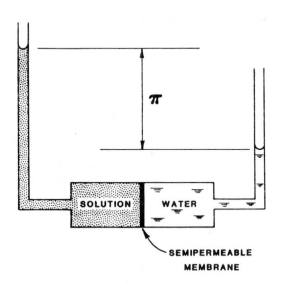

Figure 9-1. Osmotic pressure.

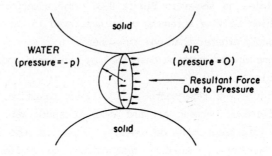

Figure 9-2. Idealized air-water interface (after McWhorter and Sunada).

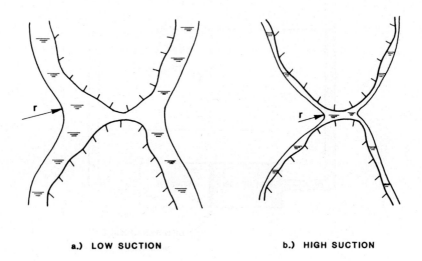

a.) LOW SUCTION   b.) HIGH SUCTION

Figure 9-3. Effect of suction on water content.

Total suction

    Total suction is the sum of the matric and osmotic suction of the soil.   For most practical applications in geotechnical engineering, the range of water contents of the soil is such that the adsorbed cations are generally fully hydrated and osmotic forces are fairly constant. Consequently, within this range, significant changes in osmotic suction do not occur [9-7].   Therefore, within the range of water contents encountered [9-8] in most practical problems, the changes in total suction that occur are those that are due only to changes in matric suction.   That is,

$$\Delta h = \Delta h_o + \Delta h_c = \Delta h_c$$

or approximately $\Delta h = \Delta h_c$

    This indicates that the stress state variable $h_c = (u_a - u_w)$ is a valid stress state variable to define soil behavior.   It also suggests that for very dry soils wherein the osmotic suction is not fully satisfied, the total suction may be the appropriate stress state variable to describe soil behavior.

MEASUREMENT OF SOIL SUCTION

    The methods of measuring soil suction are usually determined by the magnitude of suction to be measured, the characteristics of the soil, and the available equipment.   Measurement can be made using both direct and indirect measuring techniques.

    Many units are used to express soil suction in order to clarify the relationship.   The following table is presented:

Table 9-1   Soil suction unit conversion

Pf	Atmosphere Pressure (cm)	Bar	psi
1.0	10^1	9.8×10^{-3}	1.422×10^{-1}
2.0	10^2	9.8×10^{-2}	1.422
3.0	10^3	9.8×10^{-1}	1.422×10
4.0	10^4	9.8	1.422×10^2
5.0	10^5	9.8×10	1.422×10^3
6.0	10^6	9.8×10^2	1.422×10^4

Direct method

The principle of direct measurement is to create a pressure differential between the soil pore water and the atmosphere, causing the pore water to increase or decrease until its suction equals the imposed air pressure. When an equilibrium is reached between the pore water content and the imposed pore air pressure, the resultant water content is unique to this soil at this suction.

The essential component of direct measurement techniques is a membrane that is used to separate the air and water phase, so that water pressure can be measured or controlled separately from the air pressure. This membrane generally consists of a fine ceramic stone or cellulose material having a small pore space that allows the movement of water and not air. The basic principle of operation is to provide for pore spaces in the membrane that are so small that the air pressure required to force a bubble of air through the stone is greater than the anticipated suction being applied.

The pressure required to force air through the stone after it has been saturated is termed either the bubbling pressure, the air entry pressure, or the displacement pressure. All three of these terms are used synonymously. Stones having a high air entry pressure are frequently referred to as high air entry stones. Ceramic materials having a variety of bubbling pressures can be obtained commercially from several sources. The higher the bubbling pressure, the finer the stones and the lower the permeability. This can have a pronounced influence on the response time of the instrumentation and should be considered in designing a testing program [9-9]. Using the above principle, the commonly used devices are the pressure plate, the suction plate, the tensiometer and the axis translation technique.

Pressure plate: By placing the specimens on the high air entry membrane, air pressure can be applied to the specimens to increase the positive or decrease the negative pore water pressure without loss of air or loss of control of the volume of water entering or leaving the specimen. Water is forced out from the soil if air pressure is sufficient to cause positive pore water pressure. Water is imbibed into the soil if the pore water pressure is negative. Applying air pressure that introduces no flow of water in or out of the specimen is denoted as the soil suction [9-6]. The pressure plate device has a useful range of 1 to 1,500 atmosphere. When dealing with very large suction values, it is more convenient to report

those values in "pF". pF = log (centimeters of water). Thus, the range of
pressure plate is 3.0 to 6.2 pF [9-10].

Suction plate: Physically, the pressure plate and suction plate devices are
similar. Instead of applying a positive pressure to the top side of the
membrane as in the pressure plate, the suction device imposes a negative
pressure to the lower side of the membrane. When the water meniscus in the
measuring system remains stationary, the water in the soil is in
equilibrium. The negative value of the applied air pressure producing this
condition is said to be equivalent to the soil suction [9-2]. The suction
plate device can be used when suction values are less than about 2.9 pF.

Tensiometer: A tensiometer is a relatively simple device for measuring soil
matric suction. It consists simply of a fine porous stone which is placed
in contact with the soil. A pressure gauge which may consist of a dial
gauge, a manometer, or an electronic transducer, is connected to the stone
on the other side to record the pressure in the water. The stone must be
maintained in a saturated state so that air passages do not develop through
it. Consequently, its bubbling pressure must be greater than the soil
suction measured. A number of tensiometers have been manufactured
commercially for use in agricultural applications. These have also been
used in applications for geotechnical engineering.

One commercially available tensiometer that has been used with
reasonable success for measurement of suctions at a number of locations is
the "Quick Draw" tensiometer. It consists of a probe about 1.5 feet long
which is inserted into a precored hole. It has a high air entry porous tip
connected to a vacuum dial gauge through a small bore capillary tube. It
also has a null adjusting knob which allows volume changes to be imposed,
thereby decreasing the time required for equalization of the tensiometer
suction with the soil suction.

Because the porous tip allows the migration of salts through the
ceramic stone, tensiometers will measure only the matric component of
suction. For most geotechnical applications, this is adequate.

Although natural soils can contain water at matric suctions much
greater than one bar, tensiometers cannot be used to measure matric suction
much above 0.9 bars.

Axis translation: In order to extend the range of suction over which
tensiometer measurements can be used, the axis translation technique can be
utilized.

The principle of axis translation technique is shown in more detail in
Figure 9-4.   A soil sample is placed in contact with a high air entry
ceramic stone, and air pressure is applied to the sample. The high bubbling
pressure of the stone prevents air from passing through it, but the water in
the soil remains in contact with the water in the stone. This also provides
connection between the water and the water measuring device on the other
side of the stone.   Consequently, the water pressure can be maintained
throughout the system at any positive value that is desired.

As the air pressure is increased, the radii of the meniscii in the pore
water must decrease to maintain equilibrium as discussed earlier (Figures
9-2 and 9-3). Water can migrate in and out of the soil through the porous
stone to maintain equilibrium.   Thus, at equilibrium the air and water
pressures will be at the values applied and a matric soil suction equals the
difference between the air and water pressures, i.e., $(u_a - u_w)$.

To illustrate this, a soil sample is shown in Figure 9-5 in contact
with a high air entry stone. In this application, the high air entry stone
is essentially a tensiometer.   For purposes of illustration, the matric
suction in the soil is assumed to be 3.0 bars.   If the air pressure is
maintained at a value of 0.0, the water in the tensiometer reads -3.0 bars
as shown in Figure 9-5(a). As discussed above, this would not be possible
to achieve because the water would cavitate at less than one bar.
Nevertheless, assuming that the water is capable of maintaining this
negative pressure, the matric suction would be

$$h_c = (u_a - u_w) = 0 - (-3.0) = 3.0 \text{ bars}$$

On the other hand, if the air pressure in the sample were maintained at 4.0
bars, the water pressure would increase.   If the water pressure were
maintained at 1.0 bar, the matric suction would be

$$h_c = (u_a - u_w) = 4.0 - 1.0 = 3.0 \text{ bars}$$

Thus, it is evident that the suction in both cases is the same.
However, the reference against which the water pressure is measured (i.e.,
the axis) has been translated from the value of 0 bars in Figure 9-5(a) to

the value of 4.0 bars in Figure 9-5(b). This method is, therefore, called the "axis translation" technique.

This method is used frequently in the laboratory to control values of matric suction for testing of unsaturated soils. For example, in a triaxial test sample a high air entry porous stone is placed at the base, the water pressure is controlled at some value of back pressure, and the air pressure is applied through the top cap. In this way, the value of $(u_a - u_w)$ can be controlled throughout the test.

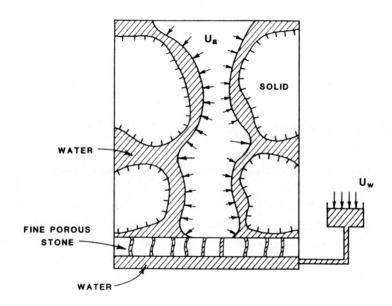

Figure 9-4. Axis translation technique.

Indirect method

The principle of the indirect method is based on thermodynamics [9-11]. Soil suction can be evaluated from the measurement of relative humidity in soils determined with a thermocouple psychrometer in the sealed chamber by filter paper. It can also be evaluated by the heat dissipation of porous material with thermal matric potential sensors.

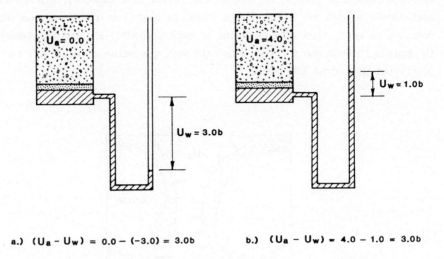

a.) $(U_a - U_w) = 0.0 - (-3.0) = 3.0b$ b.) $(U_a - U_w) = 4.0 - 1.0 = 3.0b$

Figure 9-5. Equivalency of soil suction and axis translation.

Thermocouple psychrometer: The use of thermocouple psychrometer is one of the most simple and reliable procedures for measuring soil suction [9-6].

The psychrometer consists of a semi-permeable ceramic tip which protects two very small-diameter dissimilar metal wire conductors, one usually copper and the other usually constantan. These are welded together at the tip to complete the conductor circuit [9-2].

The thermocouple psychrometer measures the relative humidity in the soil by a technique called Peltier cooling [9-12]. By causing a small direct current of about four to eight milliamperes to flow through the thermocouple junction for about 15 sec. in the correct direction, this junction cools and water condenses on it when the dew point temperature is reached. Condensation of this water inhibits further cooling of the junction, and the voltage developed between the thermocouple and reference junction is measured by a microvoltmeter.

The voltage outputs of the psychrometers are calibrated by tests with salt solutions, such as potassium chloride, that produces a given relative humidity for known concentrations [9-6].

Psychrometers are usually reliable over a range of 3.0 pF to about 4.7 pF and have been used up to about 6.2 pF. Psychrometer techniques have been advanced by various investigators including Kassiff [9-13], McKeen [9-14] and Peter & Martin [9-15].

Filter paper: The use of filter paper to measure soil suction was developed by McQueen and Miller [9-16] and by Fawcett and Collis-George [9-17]. In this technique, the filter paper is calibrated by equilibrating it with air over salt solutions of different molalities. Alternatively, the filter paper could be calibrated by use of a pressure plate and the axis translation technique as described above.

Once the paper has been calibrated, it can be used to determine soil suction as follows. A sample of the soil is placed in a closed container constructed of noncorrosive material. The filter paper is also placed in the same container and the soil sample and filter paper are allowed to equilibrate for a period of several days. During the equilibration stage, the temperature is maintained constant and is monitored. After the equilibration period, the filter paper is removed and weighed to the nearest 0.0001 g.

The filter paper method can be used over a wide range of suction up to approximately 10,000 bars [9-14]. This technique has been used in a great number of investigations of soil-water relationships and has been found to be very satisfactory for use in field investigations [9-4], [9-5], [9-14].

The principle on which the filter paper technique is based is that the soil water controls the relative humidity inside the container. The filter paper takes on or gives off water to come into equilibrium with the relative humidity inside the container as well. After sufficient time has been allowed for equilibration between the filter paper and the relative humidity in the container, the suction in the filter paper is at the same value as that in the soil.

Thermal matric potential sensor: A number of thermal matric potential sensors have been developed that work on the basis of correlating the heat dissipation in a porous ceramic with its water content. The water content, in turn, is a function of the suction.

Since the heat conductivity of a porous material is lower than that of the water, the heat dissipation in a porous material is sensitive to its water content. When a standard porous probe is inserted into a soil sample with a different pore-water tension, water passes from the area of low tension to the area of high tension. The movement of water takes place through direct capillary flow until equilibrium is reached. The rate of heat dissipation of the standard porous material, therefore, can be measured by supplying a precisely controlled amount of heat at a fixed rate at the center of the porous block and by measuring the temperature rise at the same point after a fixed period of time. The temperature rise is inversely proportional to the moisture content in the standard porous block. The measured temperature is calibrated to read matric suction.

Lee and Fredlund [9-18] performed an investigation to evaluate the performance and reliability of a similar thermal suction sensor from a different commercial source. The results indicated good performance for measuring suctions below one bar. Up to two bars the sensor exhibited lower sensitivity and its use above two bars was questionable. If the sensor was placed in the soil in a saturated condition, good agreement was observed between the results of the thermal suction sensor and the soil suction determined by other means.

THE APPLICATION OF SOIL SUCTION CONCEPT

A great deal of attention has been focused on the concept of soil suction in the recent years. Extensive research conducted by Snethen [9-4], Johnson [9-3], Fredlund [9-9] [9-19], Kassiff [9-13], McKeen [9-14] and others took giant steps in relating soil suction to the behavior of expansive soils.

As stated by Wray [9-2], most soils decrease in volume as the soil moisture content decreases and increase in volume as the moisture content increases. In terms of soil suction, soil volume decreases as the soil suction increases, and conversely, soil volume increases as soil suction is reduced. Application of this principle to unsaturated soils can result in a more realistic estimate of the shrink-swell phenomenon.

To this date, most soil suction research is concentrated on the academic. The practicing engineers are still puzzled as to the practical application of the concept of soil suction in helping to resolve everyday problems of expansive soils.

To date, the most commonly used application is in the area of heave prediction. Permeability determination as well as footing and foundation design have also entered the realm of soil suction.

The prediction of soil movement

     The procedures for predicting heave based on the suction changes in expansive soils are relatively new and are not commonly used in engineering practice.

     Richards [9-21] proposed to use moisture content-suction curves from laboratory testing to predict moisture content changes in soils. The equilibrium suction, determined from correlations between equilibrium soil suction and climatic index, is used to estimate the final moisture condition. The expansion of the soil is predicted on the assumption that the volume change is equal to the volume of water taken up by the soil. Assuming the estimated water content change is equal to the volume change of the soil, the change in volume is calculated as:

$$\frac{\Delta V}{V} = \frac{\Delta w G_s}{100 + w_i G_s}$$

where Δw = change in water content = $w_{final} - w_{initial}$

          G_s = specific gravity

          w_i = initial water content

Furthermore, by assuming equal volume changes in the vertical and horizontal directions, the heave, $\Delta H/H$ is:

$$\frac{\Delta H}{H} = \frac{1}{3}\frac{\Delta V}{V} = \frac{1}{3}\frac{(w_f - w_i)G_s}{100 + w_i G_s}$$

     The Australian method presented by Aitchison and Woodburn [9-29] requires the use of initial load and soil suction values and predicted final load and soil suction. The procedure for this approach involves the use of a modified oedometer in which applied load and soil suction can be measured. The data is plotted to provide the strain $\Delta H/H$ versus soil suction relationship for various loads. Lytton [9-22] defines the slope of this curve as the suction compression index, γ_h.

$$\gamma_h = \frac{\Delta V/V}{\log_{10} h_f/h_i}$$

where $\frac{\Delta V}{V}$ = change in volume with respect to initial volume

          h_f, h_i = final and initial suction,

Once γ_h is determined, the heave can be predicted from

$$\frac{\Delta V}{V} = \gamma_h \, \log_{10} \frac{h_f}{h_i}$$

McKeen [9-23] proposed this same procedure based on γ_h; however, he determined the initial conditions from suction tests using the filter paper procedure.

Snethen [9-6] used the soil suction–moisture content relationship for heave prediction. The data was plotted on a pF scale and a straight line approximation for the water content range of interest was represented by:

$$\log h_c^{\,o} = A - Bw$$

where $\qquad h_c^{\,o}$ = matrix soil suction without surcharge pressure

$\qquad\qquad$ A, B = constants (y intercept and slope respectively)

$\qquad\qquad$ w = water content, percent

The heave of an expansive soil profile can be estimated using the soil suction relationship as follows:

$$\frac{\Delta H}{H} = \frac{C_h}{1+e_o} \, [(A-Bw_o) - \log h_{cf} + \alpha\sigma_f)]$$

where $\qquad$ H = stratum thickness, ft.

$\qquad C_h$ = suction index, $\alpha G_s/100B$

$\qquad e_o$ = initial void ratio

$\qquad w_o$ = initial moisture content, percent

$\qquad h_{cf}$ = final matrix soil suction, tsf

$\qquad \alpha$ = compressibility factor

$\qquad \sigma_f$ = final applied pressure, tsf

The suction index, C_h, which represents the change in void ratio for a log scale change in matrix suction, is analogous to the compression index or the swell index. The compressibility factor, α, is defined as the fraction of applied pressure that is effective in altering the pore water pressure [9-24]. The compressibility factor is 1 for saturated soils and 0 for incompressible soils. When there exists no data for determining α, it can be estimated from the plasticity index by Lytton, [9-10].

$$PI \; < \; 5, \quad \alpha = 0$$

$$5 < PI \; < \; 40, \quad \alpha = 0.0275 \; PI \; - \; 0.125$$

$$PI \; > \; 40, \quad \alpha = 1$$

The above equations provide predictions of in situ volume change of a soil stratum with respect to field conditions. Vertical rise at the ground surface may be estimated by summing the volume change in each stratum in the soil profile.

Active zone

The soil suction concept provides a reliable estimation of the anticipated volume change [9-25]. The measurement of soil suction using thermocouple psychrometers is a simple, inexpensive, accurate and reliable procedure, which can be easily implemented. Soil suction tests can be used for estimating the depth of the active zone which is useful in establishing the details of the test program and providing limits for applying the prediction technique.

The depth of the active zone or the depth of desiccation has been defined as the thickness of the layer of soil in which a moisture deficiency exists. The depth of the active zone is influenced by the soil type, soil structure, topography and climate. The depth of the active zone is generally greater than the depth of seasonal moisture variation.

The soil suction versus depth profile was used by the Federal Highway Administration to estimate the depth of the active zone. The "rules of thumb" used are:

(a) For soil suction versus depth profiles that exhibited a relatively constant value with depth at lower levels, the depth of the active zone was set at the upper depth of the constant range. As shown on the typical curve, Figure 9-6, the soil suction became constant below a depth of 7.7 feet and the depth of active zone was set at 8.0 feet.

(b) For soil suction versus depth profiles that exhibited S-shaped curves with depth, the depth of the active zone was set below the first major change in magnitude of the soil suction. As shown on the typical curve, Figure 9-7, the soil suction increased to 11.2 tsf at 6.8 feet, then decreased to 6.2 tsf at 8.7 feet, which

312   FOUNDATIONS ON EXPANSIVE SOILS

constitutes the major change in magnitude, with the depth of
active zone set at 8.0 feet.

Although the "rules of thumb" do not provide exact determinations of
the depth of active zone, they do provide reasonable estimates that are
consistent with experience.

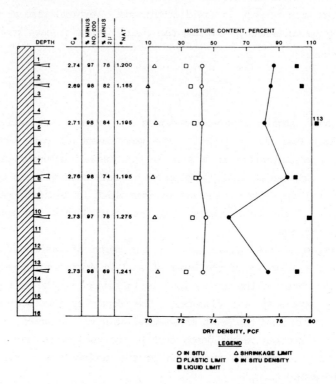

Figure 9-6. Boring log and moisture content, dry density, and soil suction
profiles.

Instability index

Soil volume change is essential for the prediction of swelling and
shrinking behavior of clays. Although volume change was initially related
to change of moisture content, soil suction is a more useful expression of
the environmental variable controlling volume change [9-26].

In order to relate vertical swell or shrinkage to change of soil
suction, Aitchison [9-27] proposed the use of Instability Index I_{p}" which is
expressed as

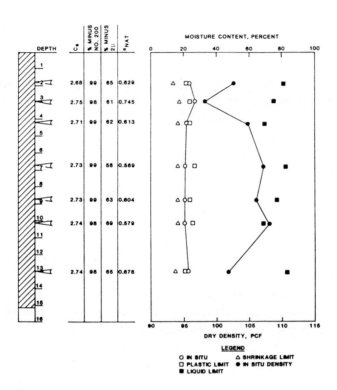

Figure 9-7. Boring log and moisture content, dry density, and soil suction
            profiles.

$$I_p" = \frac{\text{Vertical Strain}}{\text{Change of Soil Suction}}$$

Initially, Aitchison suggested in 1973 to separate the Instability
Index applied to both matrix and osmotic suction. Mitchell [9-28] applied
the Instability Index to the change of total suction.

$$I_p" = \frac{\text{Vertical Strain}}{\Delta pF}$$

Mitchell used the core shrinkage test method so that the Instability Index
can be quickly determined. With such an index determined in the general

area, it would be a simple process to determine the vertical strain by the on-site suction change.

Various methods have been developed for the measurement of the Instability Index, such as the original membrane oedometer method [9-29] and the instability cell [9-30]. The Core Shrinkage Test proposed by Fargher and Stevens [9-31] and improved by P.W. Mitchell and D.L. Avalle [9-32] involves the measurement of the linear strain versus the moisture content difference relationship ε vert/Δw as well as the moisture characteristic (the change of moisture content Δw per unit change of soil suction Δu) Δw/Δu of unconfined undisturbed samples which are allowed to dry to a moisture content above the shrinkage limit.

The Instability Index is then derived as follows:

$$I_{p''} = \frac{\varepsilon\,\text{vert}}{\Delta w} \times \frac{\Delta w}{\Delta u}$$

Predictions by this method are found to be in good agreement with the observed movement.

It should be noted that the Instability Index $I_{p''}$ is not a constant but varies with stress, soil suction, direction of change, previous history, even with the initial sample heights selected.

Permeability and Suction

The permeability or hydraulic conductivity k of unsaturated soils is the proportionality factor in Darcy's Law as applied to the viscous flow of water in soil, i.e., the flux of water per unit gradient of hydraulic potential. It has been established that this factor applies to unsaturated soils as well as saturated soils for which it was first proposed. For unsaturated soils, this factor is no longer independent of the pressure, suction or free energy of the soil water, and becomes a multi-valued function of these variables.

As the water content decreases from saturation, the large pores, which are most effective in conducting water, are the first to drain. Also, the tortuosity of the flow paths may increase, and the average properties (density and viscosity) of the soil solution may change. These factors contribute to a rapid decrease in conductivity with decreasing water content. Because of the relationship between the soil water content and the soil suction, the hydraulic conductivity may be regarded as a function of the soil suction [9-20].

Meanwhile, changes in the concentration and kinds of cationic species in the soil solution can cause a large change in the hydraulic conductivity of soils containing significant amounts of clay, especially swelling clays.

Therefore, it is possible that assessment of permeability for certain formations can be made in accordance with the measurements of in situ soil suction distribution.

Results from the capillary-moisture relationship tests are utilized in two different analyses [9-33]. Data obtained from tests performed on the proposed radon barrier cover material are used in determining an acceptable cover thickness. Other tests performed on the underlying soils and rock are based to evaluate the unsaturated hydraulic conductivity to establish potential seepage rates of contaminants and to help determine liner type and design.

The cover design is determined using a computer program which utilizes several environmental parameters, such as rainfall, temperature, evaporation, geographic location, and sample parameters such as radon diffusion coefficiencies, index properties and data from the capillary moisture tests, particularly the long-term water content. The long-term water content is obtained from capillary-moisture test results at the 15-bar pressure. It is assumed that water contents in the compacted cover materials will not decrease below this value over the life of the cover. Specimens are remolded at the long-term water content and tests for radon diffusion are performed to establish the cover thickness. It should be noted that a change in the long-term moisture content of plus or minus one to two percent could result in a change in the cover thickness of plus or minus one to four feet. This could result in millions of dollars in cost savings or cost overruns to the project.

The unsaturated hydaulic conductivity is determined from the capillary moisture relationship tests. The most common method is to calculate the conductivity by using either modified Milling and Quirk [9-34] or Campbell [9-35] methods, depending on the sample's standard properties. These methods of determining the unsaturated "k" are based on several empirical formula and some assumptions, but are the best to date. One limitation of the above methods is that they assume that the porous matrix will not undergo any change in its structural geometry as it changes from a full saturated condition to a partially saturated state; however, this will usually occur in an expansive soil.

Another parameter needed in determining the unsaturated hydraulic conductivity is the suction versus water content relationship or the capillary-moisture relationship. These methods have been updated and are published in ASTM test specifications under methods D-2325 and D-3152.

REFERENCES

[9-1]   Hilf, J.W., "An Investigation of Pore-Water Pressure in Compacted
        Cohesive Soils," Technical Memorandum No. 654. U.S. Dept. of
        Interior, Bureau of Reclamation, Design and Construction Division,
        Denver, Colorado, 1956.

[9-2]   Wray, W.K., "The Principle of Soil Suction and its Geotechnical
        Engineering Applications," 5th International Conference on Expansive
        Soils, Australia, 1984.

[9-3]   Johnson, L.D., "Influence of Suction on Heave of Expansive Soils,"
        U.S. Army Engineer, Waterway Experimental Station, 1973.

[9-4]   Snethen, Donald R. and Johnson, Lawrence D., "Evaluation of Soil
        Suction from Filter Paper," U.S. Army Engineer, Waterway
        Experimental Station, 1980.

[9-5]   Hamberg, D.J., and Nelson, J.D., "Prediction of Floor Slab Heave,"
        5th International Conference on Expansive Soils, Australia, 1984.

[9-6]   Snethen, D.R., "Characterization of Expansive Soils Using Soil
        Suction Data," Proceedings of the 4th International Conference on
        Expansive Soils, Vol. 1, Denver, Colorado, 1980.

[9-7]   McWhorter, David, and Sunda, Daniel, "Ground-Water Hydrology and
        Hydraulics," Water Resources Publications, Fort Collins, Colorado,
        1977.

[9-8]   Krahn, J., and Fredlund, D.G., "On Total, Matric and Osmsotic
        Suction," Soil Science Vol. 114, No. 5, 1972.

[9-9]   Fredlund, D.G. and Morgenstern, N.R., "Pore Pressure Response Below
        High Air Entry Discs," Proceedings of the Third International
        Conference on Expansive Soils, Vol. 1, Haifa, Israel, 1973.

[9-10] Lytton, R.L., "Theory of Moisture Movement in Expansive Clays," The
        Center for Highway Research Report No. 118-1, the University of
        Texas at Austin, Austin, Texas, 1969.

[9-11] Aitchison, G.D., "Moisture Equilibria and Moisture Changes in Soils
        Beneath Covered Areas," Australia, 1965.

[9-12] Spanner, D.C., "The Peltier Effect and Its Use in the Measurement of
        Suction Pressure," Journal of Experimental Botany, Vol. 2, 1951.

[9-13] Baker, R., Kassiff, G. and Levy, A., "Experience with a Psychrometer
        Technique," Proceedings of the Third International Research and
        Engineering Conference on Expansive Soils, Haifa, Israel, 1973.

[9-14] McKeen, R.G., "Field Studies of Airport Pavements on Expansive Clay," Proceedings of the 4th International Conference on Expansive Soils, Vol. 1, Denver, Colorado, 1980.

[9-15] Peter, P. and Martin, R., "A Simple Psychrometer for Routine Determinations of Total Suction in Expansive Soil," Third International Conference on Expansive Soils, Haifa, Israel, 1973.

[9-16] McQueen, I.S. and Miller, R.F., "Calibration and Evaluation of a Wide-Range Gravimetric Method for Measuring Moisture Stress," Soil Science, Vol. 106, No. 3, 1968.

[9-17] Fawcett, R.C. and Collis-George, N., "A Filter Paper Method of Determining the Moisture Characteristics of Soil," Australian Journal of Experimental Agriculture and Animal Husbandry, Vol. 7, 1967.

[9-18] Lee, R.K.C. and Fredlund, D.G., "Measurement of Soil Suction Using the MCS 6000 Sensor," Fifth International Conference on Expansive Soils, Adelaide, South Australia, 1984.

[9-19] Fredlund, D.G., "Prediction of Ground Movements in Swelling Clays," 1983.

[9-20] Klute, A. and Dirksen, C., "Hydraulic Conductivity and Diffusivity Laboratory Methods," 1985.

[9-21] Richards, B.G., "Moisture Flow and Equilibria in Unsaturated Soils for Shallow Foundations," ASTM, Symp. on Permeability and Capillarity, ASTM, Spec. Publ. 417, 1967.

[9-22] Lytton, R.L., "The Characterization of Expansive Soils in Engineering," Presentation at the Symposium on Water Movement and Equilibrium in Swelling Soils, American Geophysical Union, San Francisco, California, December 1977.

[9-23] McKeen, R.G., "Design and Construction of Airport Pavements on Expansive Soils," FAA-RD-76-66, Federal Aviation Administration, Washington, D.C., June 1976.

[9-24] Johnson, L.D., "Evaluation of Laboratory Suction Tests for Prediction of Heave in Foundation Soils," Technical Report S-77-7, August 1977, U.S. Army Engineer Waterways Experiment Station, CE, Vicksburg, Miss.

[9-25] U.S. Dept. of Commerce, National Technical Information Service, PB 80-139660, "Technical Guidelines for Expansive Soils in Highway Subgrades," 1979.

[9-26] Richards, B.G., Peter, P. and Martin, R., "The Determination of Volume Change Properties in Expansive Soils," Fifth International Conference on Expansive Soils, Adelaide, South Australia, 1984.

[9-27] Aitchison, G.D., Peter, P. and Martin, R., "The Instability Indices Ipm and Ips in Expansive Soils," Third International Conference on Expansive Soils, Haifa, Israel, 1973.

[9-28] Mitchell, P.W., "The Structural Analysis of Footings on Expansive Soil," Adelaide, South Australia, 1979.

[9-29] Aitchison and Woodburn, J.A., "Soil Suction in Foundation Design," Proceedings of the Seventh International Conference on Soil Mechanics and Foundation Engineering, Vol. 2, Mexico City, 1969.

[9-30] Pile, K.C. and McInnes, D.B., "Laboratory Technology for Measuring Properties of Expansive Clays," Fifth International Conference on Expansive Soils, Adelaide, South Australia, 1984.

[9-31] Fargher, P.J. and Stevens, R.L., "Notes on the Design Assumptions and Methods for Grillage Raft Footings," Hosking, Fargher and Oborn Pty, Ltd., Adelaide, May, 1973.

[9-32] Mitchell, P.W. and Avalle, D.L., "A Technique to Predict Expansive Soil Movements," Fifth International Conference on Expansive Soils, Adelaide, South Australia, 1984.

[9-33] Criley, Ken, Unpublished Research Data from Chen & Associates, 1987.

[9-34] Milling, R.V. and Quirk, J.P., "Formation Factors and Permeability Equations," 1964, Nature 202.

[9-35] Campbell, G.S., "A Simple Method for Determining Unsaturated Conductivity from Moisture Retention Data," Soil Science, Vol. 17, June 1964.

Chapter 10

INVESTIGATION OF FOUNDATION MOVEMENT

INTRODUCTION

Investigating the cause of foundation movement of an existing building and prescribing remedial measures require careful field investigation, exhaustive laboratory testing, and many years of experience. In some respects, this is similar to the treatment of a patient. Inquiry of the patient's medical record, a physical examination, and a laboratory diagnosis will be necessary to diagnose the cause of the sickness. Prescription and treatment will be relatively simple once the cause of illness has been determined. As in the case of a doctor, no examination and testing can replace experience, and experience can only be obtained by trial and error.

In the past 30 years, the author has had the opportunity to study more than 1,500 cases of cracked buildings mostly in the states of Colorado and Wyoming. These cases include residences, school buildings, offices, warehouses, swimming pools, apartment buildings, religious structures, and pavements. Most of the cracked buildings were the result of foundation movement caused by swelling soils.

HISTORY STUDY

The first step in the investigation of a building is to obtain complete information pertaining to the building. Unfortunately, such information is oftentimes absent and it is necessary to uncover much of the required information by soil exploration.

Foundation information

Effort should be made to obtain the existing foundation information relative to the soil. For buildings erected before 1960, such information is generally sketchy. Soil tests on individual sites have become a requirement after 1960. From the soil test data, it will be possible to determine the following:
1. Type of foundation,
2. Design criteria

3.   Water table condition,

4.   Type of foundation soils,

5.   Moisture content of foundation soils, and

6.   Swelling potential of foundation soils.

Sometimes, the subsoil investigation is not conducted for a specific building but for a general area. In such case, the subsoil information has only limited use. Care should be exercised to locate the building under investigation to the nearest test hole, so that it is possible to determine as closely as possible the subsoil condition beneath the building.

The above soil test data can be invaluable toward finding the cause of structural movement. The second step is to check the foundation plan. Again, such information may not be available, either because the drawing is lost or the concerned party does not want to produce it. The foundation plan will reveal if the recommendations given in the soil report have been followed. These are:

1.   The dead load pressure exerted on the footings or piers,

2.   The size of footings or piers,

3.   The length of the piers,

4.   Pier reinforcement, expansion joint, dowel bars, underslab gravel, and other details, and

5.   Subdrainage system.

If the above information is available, the investigation will be greatly simplified. This is similar to the case in which the complete medical record of a patient is at the disposal of the examining doctor. It is also necessary to examine the qualifications of the designer, whether the design is made by a registered professional engineer or by the contractor.

If the above information is not available, it will be necessary to expose the foundation system by excavation. In the case of a nonbasement building, excavation can be easily made outside of the building adjacent to the grade beam. In the case of basement construction, it will be necessary to break the concrete slab to reach the foundation. It would be a difficult job to expose the entire length of the pier, but many times it is advisable to examine the pier to ascertain a problem such as uplift.

Logs kept by the driller are sometimes available. In such cases, a complete information of the pier system will be apparent. This will also provide information on the depth of penetration into bedrock, for both interior and exterior piers, as well as the water table condition.

Movement data

Effort should be made to obtain chronological data on the building movement, items such as when the building was completed, when the first occupant moved in, and when the first crack appeared. All information obtained from the owner should be carefully scrutinized for its validity. If the owner intends to sue the builder to recover his damages, he tends to exaggerate his findings. However, with careful interrogation and keen observation, the actual story can be revealed.

When examining the exterior of the building, it is helpful to determine the lawn watering practice, the setting of the automatic sprinkling system, and the condition of the backfill. Most owners deny excessive irrigation of the lawn and flower beds.

In the interior of the building, primary information can be obtained in the basement area. Water marks or efflorescence on the wall usually tell the story of seepage water. A complete record on seepage water should be obtained: the first appearance of water in the basement, the location of seepage, the amount of observed water, and whether seepage has taken place after heavy precipitation.

Also important is the performance of utility lines. Has there been plumbing difficulty experienced in the past years? Has the floor drain been plugged? In one instance, investigation revealed that the interior house sewer was never connected to the street sewer, but emptied into the underslab soils. The defect was not discovered until an odor was detected in the basement. In another case, the basement shower drain was not connected to the sewer line. For years, the error remained undetected until the crawl space was entered and an excessive wetting condition discovered.

It is not always possible to determine the site conditions during construction, but if such information is secured by an observant owner, it can unlock many movement puzzles. There are instances where the soils were flooded during construction, and heaving movement took place even before the building was completed. Investigation of a partially completed house revealed that the basement was covered with more than two feet of snow which the contractor had failed to remove before enclosing the structure. For drilled pier foundations, it is dangerous to allow the surrounding soils to become wetted before the application of dead-load pressure. Pier uplift can begin before the placing of the foundation concrete.

In the information gathering process, all hearsay evidence should be screened. Stories such as an underground river running under the structure, the building is sliding downhill, the bentonite in the soil has pulled the building apart and others should be dismissed as hearsay by an experienced engineer, and only substantive evidence considered.

DISTRESS STUDY

The first sign of foundation movement for structures founded on expansive soils is the cracking of the floor slab. This is generally followed by doors binding, windows sticking, and cracks appearing in the exterior and interior walls and even in the ceiling.

Crack pattern

Foundation movements are reflected as cracks. Cracks caused by swelling soils have the same general pattern as settlement cracks, although swelling cracks are generally wide at the top and narrow at the bottom. The same crack patterns can develop from settlement. However, in the most severe settlement cases, diagonal cracks are usually associated with a series of horizontal cracks as shown on Figure 10-1.

Figure 10-1. Typical settlement cracks; crack pattern varies from horizontal to diagonal which is quite different from heaving cracks.

It is not always true that foundation movement of a specific portion of a structure is responsible for certain cracks appearing in the immediate vicinity of that movement. The structural arrangement of a building, especially that of a house, is complex. Movement of one portion of the building can cause cracks to appear at the opposite end of the building. It is always prudent to explain the cause of movement in a general sense and treat and study the movement as a unit. The following crack analysis can serve as a guide:

1. Diagonal cracks below exterior windows or above exterior doors generally indicate footing or drilled pier foundation movement.

2. If such cracks appear only in the exterior brick course but not on the interior dry wall, the cracks can be caused by exterior patio slab heaving.

3. Hairline cracks appearing above interior doors and closets could be caused by plaster shrinkage or timber shrinkage and not necessarily foundation movement.

4. Vertical cracks below the I-beam in the basement concrete wall can be caused by the lifting of the I-beam, resulting in tension cracks as shown in Figure 10-2.

5. Separation of the window frame from the brick course as shown on Figure 10-3 generally indicates differential heaving. Such movement has a strong resemblance to lateral movement. Actually, almost all lateral separation is caused by differential heaving.

Stress build-up

Movement of interior structural members can result in stress build-up in the structure. The most common instance is the uplift of the I-beam caused by the uplift of steel pipe columns. When the I-beam lifts, the joist system in the upper level is disturbed, doors stick and closets cannot be opened. The owner generally planes the door only to find that it fails to open and close again after a period of time. The I-beam in the basement is commonly supported by two to three steel pipe columns. When the one pipe column foundation heaves, the other pipe column is usually rendered idle and can be shaken loose by hand.

Pipe columns are provided with a screw jack at the top. The situation can be corrected, at least temporarily, by lowering the screw jack and releveling the I-beam. The doors are then able to be opened and closed freely again.

Stress build-up caused by slab bearing partition walls has been discussed under "Slabs on Expansive Soils."

Owners sometimes report that the cracks in their buildings are subject to opening and closing and attempt to correlate the movement to seasonal climate change. Such phenomenon is common in regions such as San Antonio and Dallas where shrinkage plays an important role in foundation movement. The subsoils undergo a cycle of drying and wetting, and cracks open and close following the cycle.

In areas where swelling is the predominant movement, the opening and closing of the cracks are caused by the shifting of the location of stress concentration. When a new crack appears, the stress distribution is altered, which will temporarily close an old crack. Careful observation will indicate that the total number of cracks appearing in a building is constantly increasing and seldom decreasing.

INVESTIGATION

Subsoils

To clearly define the cause of foundation movement and to recommend remedial measures, it is necessary to determine the subsoil conditions and

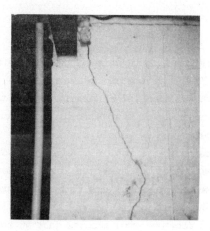

Figure 10-2. Vertical cracks beneath the beam pocket caused by lifting
              of I-beam.

Figure 10-3. Separation of window frame from brick course.

water table. Test holes should be drilled adjacent to the building and sufficient samples should be taken for the determination of the swelling characteristics and moisture content of the soil. At least one test hole should be drilled remote from the structure and in an area unaffected by building construction. The physical characteristics of the soils obtained from the adjacent and remote test holes can be compared.

It should be noted that for a building with cracking, the soils immediately below the foundation level generally have been wetted excessively. Laboratory testing will invariably show a low swell potential. However, careful testing can reveal that the material possesses a high swelling pressure. Sometimes, the actual swelling characteristics of the soil can only be revealed by air drying the soil sample and then subjecting it to wetting. In any event, samples obtained from areas unaffected by building construction should give information relative to the soil behavior at the time of construction.

The moisture content as well as the dry density of all soil samples should be determined. If possible, the moisture content should be carefully compared with the moisture content of the soil prior to building construction. In the course of nearly 1,500 cases investigated, the moisture content beneath the building area had increased. The magnitude of increase ranged from two to eight percent.

Survey

To most structural engineers, the first order of investigation of a cracked building is a survey to determine which part of the building has moved and the magnitude of movement. Such a survey is based on the following assumptions:

1. That the bench mark will not move. The bench mark chosen can be a telephone pole or top of fire hydrant.
2. That the building is constructed perfectly level. That brick courses are laid level. Consequently, it is not unusual that the survey indicates that one end of the building is as much as six inches higher than the other end.
3. That no consideration is given on the seasonal movement of the building due to temperature and moisture change.
4. That personal equation on the series of survey has not been considered.
5. In many instances, the survey circuit was not even closed.

The author questions seriously the value of most movement surveys, unless they are conducted in a professional manner and monitored and analyzed by a geotechnical engineer.

It is believed that seasonal moisture content fluctuations can result in heaving and settling of the structure. The cyclic up and down movement is believed to occur in phases but lags behind rainfall. Theoretically, shrinkage can result in settlement, but there is very little evidence that there is appreciable downward movement under covered areas in a building. Seasonal fluctuations in moisture content along the edges of highway pavement or parking areas can be expected, but at the central portion of a covered area, shrinkage seldom takes place even under a prolonged arid climate.

By referring to a reliable bench mark such as that suggested in Figure 10-4, and conducting a survey at intervals of once a month, the movement of the building can thus be monitored. The survey must be conducted with control points carefully selected. Such an undertaking is costly and time consuming. Only in the course of an important structure is such a survey warranted.

Test pits

The only positive method of determining the subsoil condition and construction details in the foundation system is by opening test pits. If

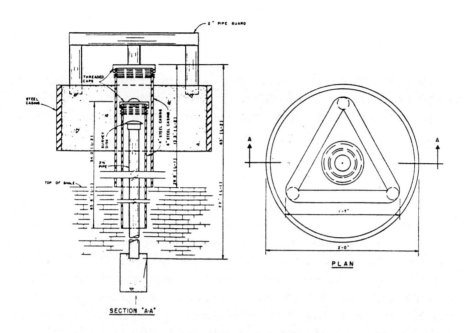

Figure 10-4. Detail of bench mark setting. (After Corps of Engineers).

the building has a crawl space, the investigation procedure can be greatly
simplified.   Pits should be opened adjacent to the grade beam and next to
the interior supports.   By investigating the test pit, the following can be
revealed:

1.   The foundation system,

2.   Condition of the air space beneath the grade beam (drilled pier
      foundation),

3.   Condition of the top of pier, the presence of mushrooms,

4.   The moisture content of the soil (soil samples should be taken every
      twelve inches to a depth of at least five feet).

5.   The presence of underslab gravel, gradation of gravel and thickness of
      the gravel layer, and

6.   Condition of the concrete slab, type of reinforcement.

Wherever possible, at least one pier should be uncovered for its entire depth and examined for possible tension cracks or voids at the bottom of the pier.

CAUSE OF MOVEMENT

When the investigation and study outlined above have been completed, the cause of movement of the building can then be determined. Obviously, no movement will take place in an expansive soil area unless the foundation soil becomes wetted excessively. Therefore, the source of moisture must be determined.

Foundation design

Foundation designs made before 1960 are based upon soil reports written with a limited knowledge of swelling soil problems and their solutions. Consequently, it is not surprising to find that the criteria established in these early soil reports are not sufficient to cope with the complexity of the swelling potential of the soil. The usual shortcomings of the given criteria are insufficient dead-load pressure exerted on a pier or pad foundation system, insufficient pier length, and lack of reinforcement.

A common design defect is the use of more supports than necessary beneath the I-beam. Dead-load pressure exerted on the interior piers is, therefore, very low. These piers are subject to uplift. A remedial measure would be to use a heavier beam section with fewer supports.

The traditional design is to dowel the exterior patio slabs into the grade beams with dowel bars. This procedure has caused a great number of heaving problems as explained previously in Chapter 6 under "Slabs on Expansive Soils."

Despite the possible deficiency of a drilled pier or footing system, a properly engineered foundation system suffers only minor distress even under the most adverse conditions. Most buildings that suffer severe damage are designed and constructed by contractors without the benefit of a soil or structural engineer.

Some contractors take the matter entirely into their own hands and prewet the foundation, puddle the backfill, reinforce the footings instead of the foundation walls, drill oversized piers and expect a stable building. Unfortunately, or fortunately, these buildings may remain in good condition for years and give an excuse to the builder to continue this undesirable practice.

Construction

Legally, if the contractor follows every detail specified by his consultants—the structural and soil engineers—then his liability in the event of future building damage will be greatly reduced. Obviously, the contractor is the first target of the owner in a lawsuit filed for negligence.

In the court, every effort will be made to prove the contractor's negligence. For instance, the absence of slab reinforcement should not be important with respect to slab cracking; however, in the court, this appears as a glaring mistake on the part of the contractor for not following the design.

Most contractors attempt to do a good job in their building. Details such as the separation of slabs from bearing walls, the use of a dowel bar to connect patio slabs with grade beams, the absence of void spaces in the slab bearing partition walls and others are the results of incomplete specifications and the ignorance of the contractor on the technique of building in expansive soil areas rather than purposely omitting the details. Unfortunately, in the court of justice, no differentiation can be made between ignorance and intentional errors.

There are exceptions in which the contractor abuses every rule of good construction practice in the foundation construction and expects to go undetected because the covered foundation excavation will not be questioned after the building is completed. In one large apartment complex, the piers were drilled off center from the grade beams, some having missed the grade beams completely. Piers were bottomed on the upper soils instead of the bedrock, resulting in chimney tilting. Some of the construction defects are shown on Figure 10-5.

Drainage

Water entering foundation soils can be from one or more of the following sources:
1. Rise of ground water or development of perched water,
2. Poor surface drainage causing surface water to enter through backfill into the foundation soils,

Figure 10-5. Timber supports used to correct the missed piers.

3.   Breakage of utility lines.

A perched water table condition cannot be foreseen at the time of construction and is usually not mentioned in the soil report; therefore, a subdrain system is generally not specified in the design.   The responsibility of damage caused by perched water is difficult to define. When there is a possibility of perched water, a subdrain system should be provided.   To provide an effective subdrain system against perched water condition involves considerable cost.

Almost 90 percent of the water that enters foundation soils is derived from surface water.   As discussed in Chapter 7, inadequate slope around the building, loose backfill, improper location of the sprinkling system, and shrubs and flower beds planted adjacent to the building are all potential causes of wetting of foundation soils.

If civil action is instigated resulting from foundation problems, the strongest defense the contractor possesses against the owner's suit is that the owner has not provided proper drainage around the building.   An alert developer issues each homeowner a manual on the care of drainage around the house.   In the event of a complaint, the builder can then have recourse back to the owner.

While it is true that poor drainage introduces water into the foundation soils and causes heaving, the role of drainage in the cracked building has been exaggerated.   Moisture content in foundation soils can increase substantially to allow heaving even if the drainage around the house is in excellent condition.

When investigating a cracked building, the most obvious defect is that of improper drainage around the building. Consequently, many investigators contend that by correcting the surface drainage, the problem is solved. Usually, this is far from the real cause of foundation movement.

When a soil engineer provides the foundation design criteria, he should design for saturated soil conditions.   In the field, he selects the soil with the highest swell potential for testing.   In the laboratory, he saturates the soil sample to determine the maximum swell potential.   His design criteria are actually based on the worst possible conditions. Precautions given on drainage are only an added factor of safety.

Unfortunately, the art of coping with present-day expansive soil problems is far from complete and the soil engineer can only hope that his recommendations are carried out in full, thus minimizing any possible future damage.

REMEDIAL CONSTRUCTION

In the past ten years in the Rocky Mountain Region, probably several thousand houses, commercial buildings, apartment complexes, school buildings and other structures have suffered damage due to expansive soils.   Most of these structures are founded with piers drilled into bedrock.   It is interesting to note that the most severe damage to structures has been associated with a perched water condition.

An analysis of building distress can add a great deal of knowledge towards the behavior of drilled pier foundations.

Symptoms and causes

The examination of a cracked building is somewhat similar to the examination of a patient. An experienced doctor will be able to diagnose the symptoms of the patient and determine the cause of the problem. Some of the more common occurrences are as follows:

Pier Movement: Pier movement is generally associated with the following:
1.   Diagonal cracks below basement windows.
2.   Diagonal cracks in the brick course.
3.   Hairline diagonal cracks in the grade beams.
4.   Ceiling cracking, often associated with patio post uplift.
5.   Uplift of the I-beam in the basement, resulting in space beneath the beam pocket and the loose pipe column.

Slab Movement: Slab movement is easy to detect and is by far the most common type of movement.
1.   Heavy cracks in concrete.
2.   Water enters the basement at one time or at frequent intervals.
3.   Subdrain system operates ineffectively.
4.   Basement floor slab heaves and cracks.
5.   Water marks on the basement wall.
6.   Absence of gravel beneath the floor slab.
7.   Poor operation of the sump pump.

Slab Induced Problems: Slab movement can induce structural damage and sometimes it is difficult to tell whether the distress is slab related or directly caused by pier uplift.
1.   Poor joint between grade beam and slab.   When the slab heaves, it transmits stress to the grade beams and can sometimes result in grade beam uplift. The brick course can crack due to such uplift, while the piers can remain stable.
2.   Slab bearing elements can transmit pressure to the upper structure. Such elements include partition walls, cinderblock walls, metal frames, door frames, staircase stringers, bookcases, wall furring, furnace connection and others. In some cases, these slab-bearing elements can cause severe distortion of the structure and even affect the exterior walls.
3.   Chimney slabs when poured monolithically with the floor slab can transmit stress to the pier supporting the chimney and result in the tilting of the chimney.

Lateral Earth Pressure: The basement wall can be affected by backfill earth pressure. This is manifested by the following:

1. Movement is shown by horizontal cracks on the wall.
2. Movement is associated with excessive wetting of the backfill soils.
3. Movement is shown by the displacement of dowel bars connecting the plate at the top of the wall.
4. Movement often takes place where the wall is long and no vertical reinforcement is used.

Structural considerations

Prior to 1940, damage to structures was usually related to poor construction or foundation settlement. Little was known about swelling soils. Today, both engineers and the general public in this region are well aware of the so-called "bentonite soils," so when a problem arises, be it a residential house, a school building or a commercial structure, the first verdict is that this is caused by the bentonite soil.

Trends in present-day construction methods have caused an increase in failure due to movement [10-1]. Among the trends, the most frequently occurring are the following:

1. The use of thinner sections with lower thermal capacity. (Steel, concrete, timber, glass, masonry and others have different thermal capacities). Combining the materials in a structure without considering the compatability can often result in cracking.
2. Large units with greater movement at fewer joints. This is especially noticeable in massive masonry structures. It is difficult to differentiate between temperature cracks in brick walls and cracks from swelling soils as shown in Figure 10-6. Temperature cracks in massive brick construction usually take place at the corners of the building in the form of near vertical separation. Failure to provide sufficient or adequate expansion joints can be the cause of such cracks.
3. The use of prestressing. The creep of prestressed concrete members, such as twin-tee floors and roofs, can continue for several years. Such movements are usually not compensated for in the design. Figure 10-7 indicates a severe crack that developed in a twin-tee structural slab. The building is founded with drilled piers, and the exterior walls are in excellent condition. These cracks had nothing to do with foundation movement.
4. The curling of concrete. Curling of concrete floors has been well documented by the Portland Cement Association. It has a strong resemblance to the heaving of slabs poured on expansive soils.

Figure 10-6. Typical crack due to temperature in massive brick
             construction.

Figure 10-7. Floor cracks due to shrinkage of twin-tee topping.

5.   Deflection can move non-structural portions that are attached to or
     resting on a structure. The intersection of non-structural partition
     structural walls will crack [10-2].
6.   Horizontal movement can be caused by wind or heat of the sun and is
     sometimes mistakenly referred to as horizontal swelling soil pressure.

     It is easy to jump to the conclusion that the drilled pier system has
moved and is responsible for the structural damage. It is sometimes
impossible to prove that the damage is actually related to other aspects of
the structure.

     A multiple story bank building founded on heavily loaded piers suffered
damage in the upper story in cracking partition walls and sticking doors.
The problem was most likely caused by the deflection of the bar joist floor
system.   Instead, the structural engineer pointed a finger to pier uplift.
Months of depositions and discovery were scheduled. Both the structural and
geotechnical engineers presented their views, but the cause of movement was
not agreed upon.   The case was settled out of court with remedial
construction involving the cutting and releveling of the piers. The cost of
the remedial construction was borne by both the contractor and the
geotechnical engineer. The building continued to move, but the case was
closed.

     In a court of law, expert witnesses summoned by both the plaintiff and
the defendant can have totally opposite views and the attorneys will
continue to find loopholes in the exhibits. Judges and juries are sometimes
confused and often reach a verdict based on emotion.   The truth of the
matter may never be discovered.

     While it is true that swelling soils are probably responsible for most
of the cracking and movement of lightly loaded structures, other aspects of
foundation movement cannot and should not be ignored.

Pier replacement

     It has been a common practice recently for the structural engineer to
order the replacement of uplifted piers with new piers. Bearing in mind
that pier movement is unusual unless associated with a severe perched water
condition, it is unreasonable to suspect uplift of heavily loaded piers on
the order of 500 kips. Pointing a finger to inadequate void space and
slight mushrooming is not usually justified.   Still, many structural
engineers and all laymen believe that pier uplift is the root of structural
distress. If they cannot establish the movement of a single pier, they

dream up the theory that there is a general slide or massive uplift of soils in the general area responsible for the problem. Such a theory is difficult to deny. Try to prove this in front of a judge. The geotechnical engineer usually winds up on the losing end of the dispute.

As seen in Figure 10-8, perched water flows only in the seams and fissures of claystone. Water enters through the seams of claystone to the interface between the pier and soil and causes pier uplift. The moisture content of the claystone mass can vary greatly. When taken at the seams or fissures the moisture content can be very high, yet only a short distance away from the seams the soil can be dry.

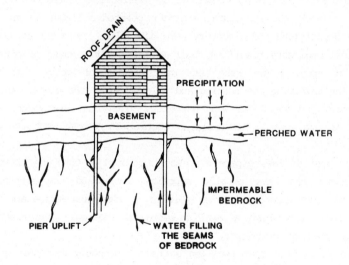

Figure 10-8. Water entered the interface between pier and soil causing uplift.

The existing piers have uplifted. In this case, they have reached almost the full amount of uplift and the future uplift movement will be small. The uplifting mechanics of the piers in this case are quite different from the heaving of the underslab soil. For underslab soil heaving, the movement is never stabilized and continues as long as moisture is not removed.

Consequently, in this case, the remedial measures should consist essentially of releveling the existing piers. This consists of jacking up the grade beam, cutting off the top of the piers and shimming the piers to achieve a level condition.

By drilling new piers, the piers penetrate into the dry claystone but also cut the water-filled seams and eventually cause the new piers to uplift. This is an undesirable undertaking. The future uplift of the new piers can damage the repaired residence and cause renewed problems. The author strongly advises against such construction.

LEGAL ASPECT

Soil engineers, particularly in the Rocky Mountain area, are well aware of the numerous litigations arising from the owner suing the builder, the architect, and especially the soil engineer, on buildings founded on expansive soils [10-3].

The mechanics of swelling soils are not fully understood, with engineering study dealing with expansive soils still in its infancy. Only in the last 20 years have soil engineers turned their attention to this very complicated problem. Although much research and study have been directed to expansive soils, the basic problem of how to deal with lightly loaded structures on expansive soils remains unsolved. A practical method on how to design a structure that will be safe against soil heaving within economical reach has yet to be found. A drilled pier foundation may be satisfactory in many instances, but it is not always a solution. The buildup of perched water or excessive increase of moisture content due to uncontrolled irrigation can often cause pier uplift and result in irreparable damage. The owner has little idea of the problems involved in the making of a rational design. He assumes that if a soil engineer is paid to design the structure, it is his responsibility to turn out a perfect foundation system, and when the first sign of cracking appears, the soil engineer is the one to be blamed, and a lawsuit follows.

In a way, one can compare cancer research with the expansive soil problem. It appears that expansive soil is the cancer in soil mechanics. Similar to cancer, the seriousness of the problem has only been discovered in recent years, with the gravity of the problem increasing at an alarming rate. Only 15 to 20 years ago, swelling soil did not present a threat to the construction industry. It has now become a major concern.

In the last decade, a large amount of money, probably reaching a billion dollars, has been poured into cancer research, yet a complete cure for cancer is still very remote. When a cancer patient dies, the doctor does not get sued for failure to cure the patient, as he has treated the patient to the best of his knowledge. Yet, when a building founded on expansive soils cracks, the engineer gets sued. The court does not give allowance to the fact that there is no positive solution for a lightly loaded structure founded on expansive soils. It does not seem to realize that the engineer has done everything possible for his client based on present-day knowledge of expansive soils. Today, only a trickle of money has been directed from the federal government toward expansive soil research. Most of the research on this subject has been carried out by private consultants with limited financial resources.

The chance of a cancer patient recovering depends a great deal on early detection of the symptoms, the cooperation of the patient with the doctor, the physical condition of the patient and other reasons. Such is the expansive soil problem. The chance of a building free of problems depends on good foundation recommendations, a contractor who follows closely the recommendations of the soil and structural engineer, and the owner's diligent maintenance to keep excessive moisture away from the building. It is grossly unfair in today's justice that a soil engineer should be blamed for what happens to the structure when he has little control over its construction and when knowledge of this field is still in darkness.

Responsibility

In the field of construction, it is difficult, if not impossible, to pinpoint the cause of damage. It is even more difficult to define responsibility. Most of the damage incurred in buildings founded on expansive soils in the Rocky Mountain area arises from slab-on-ground construction. The soil engineer warns the owner that he should use a structural slab where there is no contact between the slab and soil. The owner cannot afford to construct such structure and elects to use slab-on-ground. When the slab heaves, other structural members suffer. Even though a floating slab is advocated, it is difficult to isolate slab heaving with other structural defects. To a layman and to an attorney, there is no segregation between the damage caused by slab heaving and general distress.

The house is sliding down the hill, complains the owner, and his attorney agrees. Exhibits shown to the court indicate the ugly cracks, magnified several times. The owner swears on a stack of Bibles that he has followed every word the soil engineer has recommended, that he has never watered his lawn more than once a week, and that the shrubs and plants immediately against the house never require watering. He conveniently forgets to tell the court that he left the garden hose running in his yard when he left for vacation.

Members of the jury who have any knowledge of home building, such as realtors, contractors, engineers, architects and even insurance agents, have long been disqualified by the plaintiff's attorney. The owner wants full compensation for the damage suffered on his investment plus a large sum of punitive damage. He does not care whether the money comes from the builder or the engineer. The builder claims that he has followed the specifications and drawings prepared by the engineer, and he blames the owner for excessive irrigation around the building. The engineer claims that most of the damage is a direct result of slab heaving, and he has forewarned the owner that a structural slab should be used. During the deposition, the structural engineer, the landscape architects, and even the plumbers and carpenters are involved.

Act of God

The drilled pier foundation is able to provide an adequate foundation system for lightly loaded structures in expansive soils by providing sufficient dead load pressure and anchoring the pier into the zone unaffected by moisture change. However, in the case of the rise of ground water, or, more often, the development of a perched water condition, the anchoring effect of the pier is lost and pier uplift follows.

Perched water generally takes place in areas where bedrock is high. After the surrounding area has been fully developed, water from surface irrigation and precipitation is trapped on top of the bedrock and a perched water condition prevails.

Pier uplift causes great damage to the building and remedial action is difficult and costly. Such cases, when brought to the court, could generate a great deal of argument. The soil engineer claims that he has designed the system correctly and to the best of his knowledge, that the area had no high water table condition at the time of his investigation, and that the development or rise of perched water is, indeed, an act of God.

An act of God is defined legally as, "An eventuality outside human contemplation." It generally entails either the unforeseen ability by reasonable human intelligence or the absence of human agency causing the alleged damage. If a similar rise of ground water had occurred before, or could have been anticipated using modern techniques or if the damage were otherwise reasonably foreseeable, even if not probable, an "Act of God" will not serve as a defense.

The attorney for the plaintiff, therefore, argues that the engineer has not designed the structure in good faith, that he should have foreseen the possibility of a perched water condition and designed the foundation system for such occurrence, and that he has not exercised reasonable care and competence. Since the contractor has constructed the building in accordance with the design and specifications and since the structural engineer and the architect have designed the building according to the soil engineer's established criteria, therefore, the soil engineer is solely responsible for the alleged damage.

An expert witness is called for the plaintiff and defendant. The experts may differ in opinion depending on experience, training and education, and above all, their court experience. The jury cannot understand why the soil engineer cannot design a building properly so that the ugly and continuous movement of the building does not take place. The judge, after a tour of the structure, is appalled by the extent of damage. What chance does the soil engineer have, although he knows all along that this was an act of God.

The pattern of the lawsuit

Lawsuits against geotechnical engineers for structural damage on expansive soils in the Rocky Mountain area probably have reached several hundred cases each year. It appears to have followed a certain pattern. The following is a typical case:

A typical basement in a residential custom-built house is founded on drilled piers. The foundation system was designed by a structural engineer using criteria furnished by a geotechnical engineer. Two years after occupancy, the owner found water had seeped into the basement, doors were sticking and ugly cracks were showing on the basement concrete floor. The owner contacted the contractor. The contractor claimed that he had built the house in accordance with plans and specifications provided by a registered engineer. The insurance company agreed to make cosmetic repairs but refused to perform major repairs. They claimed that the policy covers only "structural damage." Although the cracking and movement of the house was extensive, it was not structural related.

Frustrated, the owner retained an attorney. The attorney promised him that he would take the case on a contingency basis. That is, if the case is lost, the attorney will not charge the owner for his legal fees. If an award is made, the attorney would take 60% of the judgment. Named in the suit were the contractor, architect, structural engineer, subcontractors and, of course, the geotechnical engineer.

After three years of depositions, discovery, settlement conferences, etc., a trial finally took place. The jury consisted of five housewives and one salesman. All of them were homeowners. The highlights of the trial were as follows:

1. The owners claimed that they have lost the entire value of their house which was purchased eight years ago at $50,000. They asked for the full value of the house at current market price or repair of the house to its original condition. In addition, they claimed pain and suffering loss during the period the house suffered damage for $1 million.

2. During the discovery investigation, the plaintiff's engineer found an incomplete void under the grade beam, mushrooms at the top of the piers, and improper expansion joints in the floor slab. It was alleged that such defects were the cause of the existing damage. Surveys were made indicating differential movement as much as six inches. However, no reliable bench mark was established and the movement readings were based on the assumption that the house was constructed perfectly level. with the survey plotted in a scale in which the vertical is 100 times that of the horizontal, it presents a dramatic exhibit in front of the jury.

3. Water in the basement was a major issue. When the geotechnical engineer conducted the initial soil test, no ground water was found to a depth of 50 feet. Subsequent development of a perched water condition was interpreted by the attorney for the defendant as "an act of God."

4. The plaintiff's attorney engaged a professor from a small college who had no experience either in the field of expansive soils or in court. He testified that the geotechnical engineer should use modern technology, including finite analysis, modelling and X-ray defraction tests to prepare his original soil report. The following key question was asked:
    "Did the geotechnical engineer employ that degree of knowledge, skill and judgment ordinarily possessed by members of that profession, and did he perform faithfully and diligently all services undertaken as an

engineer in the manner a reasonably careful engineer would do under the same or similar circumstances?

The answer was an unqualified "No." Thus, the fate of the geotechnical engineer was sealed.

Needless to say, the outcome of the trial is that the defendants were found guilty. The share the geotechnical engineer had to pay was considerable. On top of that, he had to pay for his attorney and three years of his time in dealing with this case, probably all for the $500 fee he received for his soil test. No wonder geotechnical engineers now refuse to perform soil tests for residential houses.

Future trends

The author can see very little relief in our legal system towards the litigation of the swelling clay problem. It is important to face the issue from both the engineering point of view and the legal point of view. From the engineering point of view, we must emphasize the following:

1. The consulting engineer, the academicians and the civil service engineers must have a united front toward the issue concerning expansive soils.

2. The public, as well as the legal profession, must be educated to realize that expansive soil is a problem which cannot be fully resolved with present-day knowledge.

3. The structural engineers should cooperate fully with the geotechnical engineers. Attempts should be made to differentiate structural-caused damage from soil-caused damage.

4. Engineers should learn the art of giving depositions and testifying in court. By realizing that hindsight is 20/20, it is always possible to find apparent defects in the soil report that the attorney usually points out.

5. An all-out effort should be made to persuade the government and the industry, as well as the engineering community, to make extensive research on this greatest natural hazard that is facing us.

To avoid litigation and to encourage good engineering, it is necessary to educate the public on the problem of expansive soils. The Public Awareness Conference on Home Construction on Shrinking and Swelling Soils held in Denver in 1978 attracted large participation. Such a conference should be held more frequently. The participants should include not only engineers, but also attorneys, realtors, insurance agents, architects, and, above all, homeowners. Litigation will not solve the problem of expansive soils. On the contrary, it will impede the development and betterment of this aspect of soil mechanics.

Effort should be made in state legislation to curb frivolous suits on soil engineers in connection with expansive soils. It is also necessary for the legislators to take a hard look at the statute of limitations on structures founded on expansive soils. In most cases, ten years is too long, and probably three years would be more reasonable.

The recent movement towards tort reform, both at the federal and the state level, has somewhat improved the status of litigation against engineers. Statutes, such as a provision allowing pre-trial screening panels for lawsuits against design professionals, can greatly reduce cases against engineers.

The author is looking forward to a breakthrough in expansive soil technology before the year 2000. Only then can we avoid the 4.5 billion dollar annual loss, and engineers will no longer be haunted by liability claims.

REFERENCES

[10-1]   Rainger, Philip, "Movement Control in the Fabric of Buildings," Batsford Academic and Educational, Ltd., London, 1983.

[10-2]   Weingardt, Richard, "All Buildings Move - Design for It," Consulting Engineer, 1984.

[10-3]   Chen, F.H., "Legal Aspects of Expansive Soils," Fourth International Conference on Expansive Soils, 1980.

PART II

CASE STUDIES

DISTRESS CAUSED BY PIER UPLIFT

GENERAL

The case study is that of a school building (Fig. I-1) and is typical of pier uplift. The pier load is heavy, construction is in general up to standard, design is adequate, and crawl space construction allows the structure to be free of possible damaging effects of slab heaving. Yet damage to the building caused by uplift, before positive remedial measures were taken, was so severe that evacuation of the building was considered for reasons of safety.

HISTORY

The school was completed in 1962. It is founded with piers drilled into bedrock. The bedrock consists of essentially claystone and sandstone shale located at depths 4 to 23 feet below the ground surface. The piers were designed for an end bearing pressure of 20,000 psf and a skin friction value of 2,000 psf. The piers were also designed for a minimum dead load pressure of 15,000 psf. The piers were to penetrate the shale bedrock by at least 4 feet and only the skin friction in the bedrock was to be assumed. The pier design system was considered to be sound in view of the limited knowledge of pier design in 1960.

Shortly after completion, distress of the building was noticed. In June, 1964, the contractor was advised to repair the existing damage. In November, 1964, the piers were found to be in good condition but surface drainage had not been properly provided for and water had penetrated beneath the structure causing soil swelling.

In July 1966 various columns were jacked up to level the building and steel columns were inserted for support. In February 1970, as a safety precaution a precast panel over a doorway had to be removed. In July and December 1970 repairs were made on several piers. In June 1971 further repairs were made to a number of interior columns. An inspection in July 1971 indicated that movement was still continuing.

WING B

WING C

Figure I-1. Exterior view of school building under study.

In March 1972 the author was engaged to make a complete independent investigation into the cause of cracking and to determine the necessary remedial measures.

The school building consists of three levels. The lower level is designated as Wing C. This level has a lower floor and one story above the lower floor. The middle level, designated as Wing B North, and the upper level, designated as Wing B South, are both one story high with no lower floor. The gymnasium, cafeteria, and music hall are all one story with a high ceiling and designated as Wing D. The northern portion of Wing D has a basement locker room. This is the only portion of the entire building where slab-on-ground construction is used. The remainder of the building is crawl space type construction. Wing A is located at the west side of the building

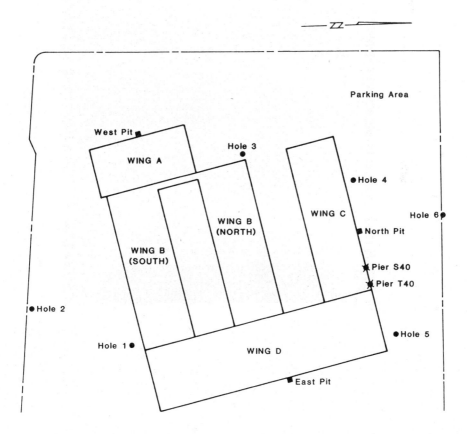

Figure I-2. Plan of school building under study.

Figure I-3.   Compression failure of pedestal placed above pier and beneath grade beam. Uplifting pressure 30,000 psf.

and is occupied by a library and administration building. The remedial construction on Wing C began in 1973 and was completed in 1974. The various wings are shown on Figure I-2.

Most of the remedial measures undertaken in the past ten years were centered at Wing C. In the crawl space under Wing C, many concrete pedestals were removed and replaced with steel H columns. Figure I-3 shows that the concrete pedestals were crushed by pier uplifting force in the same manner as concrete cylinders are crushed in the compression testing machine. Along the north wall in the crawl space, water seeped freely into the crawl space through the backfill. Water has been entering below the grade beam for a lengthy period of time. At the northwest corner, evidence was found that water has flowed in freely and has washed the soil in the crawl space forming channels. In Wing B, both north and south, the ground surface was relatively dry, but there was evidence that the soil has been wetted in the past due to infiltration of surface water.

The lower level construction is confined in Wing C. The east wall of the workshop revealed severe movement. The arts and crafts rooms also showed many areas of extensive damage. Surprisingly, the slab-on-ground portion of the locker and boiler rooms showed practically no foundation movement.

In the upper level, numerous cracks were found in the interior walls of Wings B and C. Almost every interior partition in these areas was

Figure I-4. Cracks and separation of brick from ceiling.

cracked. The ceiling had pulled away from the structural walls by as much as 3 inches at the north end of the north-south corridor below Wing D and the rest of the building as shown on Figure I-4.

Exterior cracks were found in the grade beams and brick courses, particularly on the north wall of Wing C. Hairline cracks were found in the reinforced concrete beams in the crawl spaces, particularly in Wing C.

INVESTIGATION

Swelling potential

Six test holes and two test pits were excavated in 1972 at the locations shown on Figure I-2 and undisturbed samples taken from the test holes. Swell tests conducted on samples taken from test holes located at the south side and west side of the building indicate that the expansion is about 1 to 2 percent and the swelling pressure is about 3,000 to 4,000 psf. Since the test holes were drilled adjacent to the building, it is likely that due to excessive wetting condition, the soil has already swelled to its maximum limit. Consequently, the swell tests cannot reveal the initial soil condition.

Swell tests performed on undisturbed samples taken from test holes drilled outside of the building area present a different condition. The upper clays swell about 6 percent with a swelling pressure as high as 25,000 psf.

The typical dry clays found near the crawl space, after remolding and upon subsequent wetting, exhibit high swell potential as shown on Figure I-5. The swell characteristic represents the actual condition of the subsoil at the time the building was constructed.

Assuming the swelling pressure of the upper clays is 15,000 psf, the following calculation will indicate the stress condition around the piers:

Data:	Pier diameter	= 30 in.
	Pier circumference	= 7.8 ft.
	Pier end area	= 4.9 sq. ft.
	Portion of pier in upper clay	= 8.0 ft.
	Portion of pier in bedrock	= 4.0 ft.

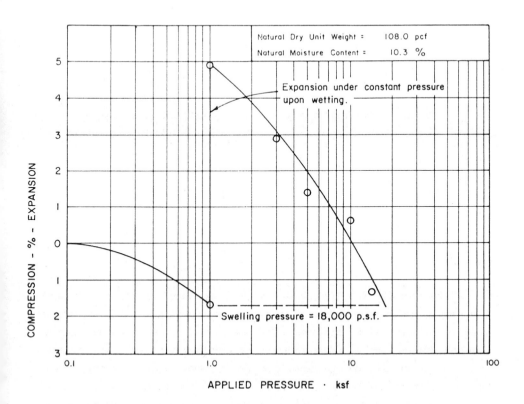

Figure I-5. Typical swell test performed on remolded samples.

                Swelling pressure of
                   upper clays   = 15,000 psf
                Portion of swelling
                   pressure responsible
                   to uplift   = 15,000 x 0.15
                                          = 2,250 psf
                Skin friction   = 2,000 psf

    then

Total uplifting force   = (Total area of pier
                                             exposed to wetting) x
                                             (Unit uplift)
                                         = 7.8 x 8 x 2,250
                                         = 140.4 kips

Total withholding force   = (Total area of pier in
                                             bedrock) x (Unit skin
                                             friction)
                                         = 7.8 x 4 x 2,000
                                         = 62.4 kips
Net uplift force   = 140.4 - 62.4 = 78.0 kips

Moisture analysis

        Three piers, S-40, T-20, and W-38 were excavated to their full depth
and samples taken to show the variation in soil properties with respect to
both depth and radial distance from the piers.   The moisture content
determined from samples taken in the test pits and test holes was compared
with the values obtained on samples taken in 1961.   Samples taken in 1972
both adjacent to the building and well away from the building were compared
to determine the nature of moisture movements.   The data are summarized as
follows:

North Wall - Upper Clays

Year	Location	Avg. moisture content, percent
1961	In test holes	17.1
1972	In test holes	21.1
1972	In test pits	28.1
1972	Along pier T-40	25.8
1972	Along pier S-40	28.9
1972	Remote from the building, Test hole 6	15.2

The data are not definitive, but suggest that a significant increase in moisture content has occurred in the soil next to the building and around the piers.

North Wall - Bedrock

Year	Location	Avg. moisture content, percent
1961	In test holes	21.0
1972	In test holes	20.3
1972	Along pier T-40	23.7
1972	Along pier S-40	23.4
1972	Remote from the building, Test hole 6	19.3

The above data shows that the moisture content of the lower bedrock has remained fairly uniform in the past 10 years, indicating that it has not been substantially wetted. The bedrock immediately adjacent to the pier appears to have slightly increased in moisture content.

East and West Walls - Upper Clay

Wall	Year	Source	Avg. moisture content, percent
West	1961	From test hole	17.7
	1972	From test hole	21.2
East	1961	From test hole	18.3
	1972	From test hole	18.9

A significant increase in moisture content has occurred for the west wall, but the change along the east wall is negligible. The bedrock actually appeared to be drier in 1972 than in 1961.

To determine if the moisture was penetrating along the walls of the piers or soaking down uniformly from the surface, moisture content samples

were taken adjacent to the walls of piers S-40 and T-40 and also 3 feet
away. The average moisture content was as follows:

Pier	Avg. moisture content, percent			
	Upper clay		Bedrock	
	At wall of pier	Three ft. away	At wall of pier	Three ft. away
S-40	28.9	20.3	23.4	18.8
T-40	25.8	21.0	23.7	21.3

The data strongly suggest that the main moisture movement is
immediately along the surface of the piers.

Pier uplift

The increased moisture content around piers S-40 and T-40 suggest that
both have been subjected to uplift. For pier T-40, it is possible that
surface water has entered along the face of this pier and may have even
reached near the bottom of the pier. Consequently, the entire pier has
lifted. For pier S-40, a 3/8-inch-wide horizontal crack was found just
above bedrock. Since excavation of the pit around the pier relieved all the
uplift forces on the side of the pier in the clay, the pier should have
gradually settled as the pit was excavated and it was theorized that the
crack must have been open by more than 3/8 inch prior to excavation.
Tension cracks developed in the pier clearly indicate that the upper soils
have exerted uplifting pressure on the upper portion of the pier, and the
portion of the pier in bedrock is withholding the pier.

The uplifting pressure exerted on the pier depends on the swelling
pressure of the surrounding soils. The uplifting force exerted on each pier
may reach as high as 200 kips. This force is sufficient to crush the
concrete pedestal formed on top of the pier. Also, when all the piers in
Wing C were exposed during the remedial construction, it was found that at
least five piers had a distinct shear failure pattern as shown on
Figure I-6.

CAUSE OF MOVEMENT

In general, the cause of movement of the building is due to the uplifting of the piers. The movement is more severe at the north side under

Figure I-6. Failure of pier by shear resulting from uplift.

Wing C. To the west of Wing D, the entire school building is connected with grade beams. Wing D is separated from the remainder of the school building with expansion joints. Consequently, at the eastern portion of the school building foundation movement is distributed throughout the system and is not conspicuous, while at the north-south corridor the entire system is separated. This explains why severe movement along the north-south corridor is noticed.

In addition to the uplifting of the piers, several other construction defects were found. Two piers in Wing C are bottomed on the upper clay instead of drilled into bedrock as shown on Figure I-7. The piers were 36 inches in diameter and 3 to 4 feet in length, rather than 12 inches in diameter and drilled into bedrock as had been designed. Since the upper clays have a maximum soil bearing value of about 3,000 psf, it is possible that settlement of these piers has taken place.

Figure I-7. Improperly placed pier. Pier length should be 20 feet
            and bearing on bedrock. Actual length only 4 feet and
            bearing on clay.

Figure I-8. Four-inch void, which has completely closed,
            beneath the grade beam.

The entire length of air space beneath the north wall in Wing C was carefully inspected. There was a minimum of air space. Remnants of cardboard used for forming the air space were found, but it appears that the air space was not properly constructed, as shown on Figure I-8. Either the air space was not formed to the specified thickness or the uplifting of the soil has closed the air space. In any event, along the north wall in Wing C, the soil has exerted uplifting pressure on the grade beam that can reach as high as 25,000 psf.

Not all the distress manifest in the building was caused by foundation movement. All the partition walls in the classrooms show cracks. The pattern of cracks indicates that the beams supporting the slabs were deflected. The cracks in the partition wall are typical distress due to the deflection and plastic flow of the long-span concrete floor beams.

It is important to isolate structural defects from foundation movement when investigating a cracked building so that the cause may be determined. In many cases, structural defects and foundation defects take place in the same structure.

REMEDIAL CONSTRUCTION

Since the cause of foundation movement and the source of moisture that entered into the foundation soils have been defined, the remedial measures should consist essentially of relieving the uplifting pressure exerted on the piers and preventing additional water from entering the foundation soils. The remedial measures consist of the following:

1. Remove all backfill around the building and replace compacted to at least 90 percent standard Proctor density at optimum moisture content. Backfill along the north wall of Wing C should consist of nonexpansive soils instead of the original soil. The adequate compaction of the backfill soil is very important to insure that any surface water will not penetrate through the backfill and into the foundation soils.

2. All void space beneath the grade beam should be re-formed to insure that there will be at least 4 inches of space between the soil and the grade beam. At the same time, care should be taken to insure that there will be no large mushrooms present on top of the piers. The air space should be formed adjacent to the sides of the piers. With the air space properly formed, the load of the building will then be exerted on the piers.

3.   In Wing C, it will be necessary to loosen or remove the soils above bedrock from around all piers. Such undertaking will have to be performed by hand inside the crawl space. The depth of loosening or removing of soil should be at least 8 feet.

4.   Drainage around the building must be improved, and should consist of the following:

   a.   Improve the drainage in the courtyard area. Remove the asphalt paving in the courtyard and replace with concrete.

   b.   Reconstruct the concrete sidewalk around the building to provide an adequate slope. Also provide an adequate expansion joint between the sidewalk and the grade beam.

   c.   Slope the ground surface around the building away from the structure to allow proper drainage.

The above remedial measures will prevent further damage to the school structure due to expansive soils. After the above remedial measures are made, the dead load pressure will be fully exerted on the piers and the upper soils will not exert uplifting pressure on the piers. By preventing water from entering the crawl space area, movement of the piers should be arrested.

The releveling procedure can be started, as follows:

1.   Carefully establish the elevation of all piers beneath the school building by referring to the established bench mark at the north side of the building.

2.   A structural engineer should be consulted to determine the appropriate new elevation of the school building commensurate with the initial construction.

3.   The piers around the exterior of the school building have lifted; however, at the central portion of the building the piers have maintained their original position. It is reasonable to lower the exterior piers and allow the interior piers to maintain their original position.

In 1972, remedial measures as recommended above were started. To facilitate construction, the entire crawl space area beneath Wing C was lowered. This not only allowed workmen to move freely in the work area but also remove at least 4 feet of soil around the piers (Figure I-9). The entire crawl space was lighted and conveyor belts installed for earth removal. Each pier was carefully examined for defects after the surrounding soil was removed. Steel rings were installed around the top of those piers that suffered shear failure. All grade beams were examined for structural

Figure I-9. Loosening of soil around the pier to eliminate
            uplifting pressure.

Figure I-10. Steel ring placed around the defective pier and
              steel girder installed to strengthen the grade beam.

strength.   Heavy steel girders were introduced to strengthen the defected beams (Figure I-10). Other remedial measures such as providing adequate air space beneath the grade beams, removing and recompacting backfill, installing sump pumps to eliminate perched water, and relocating those piers having insufficient length were performed under close supervision.

Then the operation of releveling started. The grade beams were raised with high-capacity jacks (Figure I-11), and the top of pier cut-and-shimmed with steel plates. Three or four piers were releveled in one operation. A total of 56 piers were releveled in Wing C over a period of 4 months. During the leveling operation, careful surveys were conducted to determine the vertical movement. Typical records are shown on Figure I-12.

Figure I-11. Jacking the grade beam in the releveling operation.

Four sets of major readings were taken as follows:

1.   Pier elevation before remedial construction,

2.   Pier elevation after air space beneath the grade beams was cleared and load of building concentrated on the piers,

3.   Pier elevation after the removal of soils surrounding the piers, thus partially eliminating the uplifting pressure exerted on the face of the piers, and

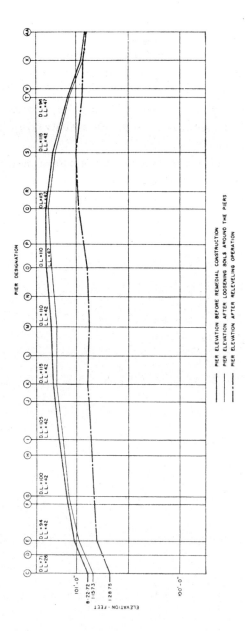

Figure I-12.   Pier settlement after various stages of remedial construction.

4.   Pier elevation after the releveling of the piers.

From Figure I-12, the effect of the various stages of remedial construction can be reflected by the settlement of the piers.

The case study of this school is a typical example of failure due to pier uplift. In addition to the construction defects, present knowledge of a drilled pier system in expansive soils calls for both adequate reinforcement of the pier to resist tension and deep penetration into bedrock to provide for anchorage. Such precautions could have resisted the uplift pressures.

Remedial construction for this school building has been confined to Wing C. After a period of 6 months, the building is still undergoing structural adjustment. Minor cracks appeared in the block wall as the result of releveling adjustment. It is expected that a stabilized condition can be achieved in the building within a year.

DISTRESS CAUSED BY IMPROPER PIER DESIGN AND CONSTRUCTION

GENERAL

This is a typical case of improper design and construction of a drilled pier foundation system. The building is a residential house located in west Denver, Colorado.

EXISTING CONDITION

Design

The residence is a split-level structure facing east, with the finished basement at the south end, crawl space under the living portion and garage at the north end. It is a brick veneer and wood frame structure with a trussed roof system and supported on piers (Figure II-1).

A subsoil investigation was made before construction. The subsoils consist of about four feet of stiff clays overlying claystone bedrock. The water table was found at a depth of seven feet below the original ground surface.

A pier foundation was recommended. The piers were designed for a maximum end pressure of 15,000 psf, a skin friction of 1,500 psf and a minimum dead load pressure of 15,000 psf. It was also recommended that the piers should be drilled at least four feet into claystone. (As claystone bedrock was practically exposed in the excavation, the length of the pier does not exceed four feet).

Distress

The house was built in 1961. Cracks appeared in the house six months after occupancy. The extent of movement began increasing steadily. A subdrainage system leading to a sump pump was later installed in the crawl space area of the house. The most severe movement took place between the crawl space area and the living room area. The separation of the crawl space from the two-story portion of the house is shown on Figure II-2. The width of the separation measures as much as 1 1/2 inches. The picture

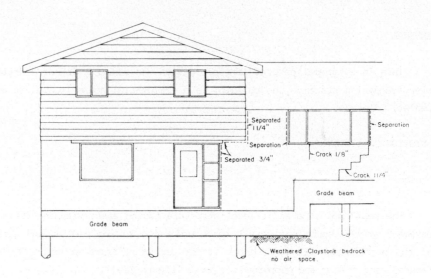

Figure II-1. Location of exterior cracks.

window above the crawl space area is also separated from the wall by as much as one inch.

Cracks also appeared at the rear of the house between the one and two-story portions. Exterior doors were jammed and the patio slab was approximately one inch lower than its original position.

In the interior of the house, severe cracks were found near the staircase leading to the basement (Figure II-3). Cracks were also found above most doors and windows, indicating severe movement. Most of the doors in the house were jammed. The I-beam which supports the upper floor appears to have moved. One of the posts in the crawl space area was loose, indicating the uplifting of the support under the I-beam. Slabs were raised.

In general, the extent of cracking in this house is considered to be very severe, and from the pattern of the cracks and the nature of the swelling of the foundation soils, severe uplifting movement of the soil beneath the foundation has taken place.

Figure II-2. Separation of living room from the
two-story portion. (See previous figure).

CAUSE OF MOVEMENT

The cause of foundation movement for this house can be summarized as
follows:
1. Undisturbed hand drive samples were taken in the crawl space area
    beneath the grade beams. Tests indicated that the weathered claystone
    possessed high swell potential. Typical test results are shown on
    Figure II-4. Figure II-4 indicates that the swelling pressure is about
    16,000 psf.
2. No air space was found beneath the grade beam near the main entrance in
    the crawl space. The total length of the portion of grade beam without
    void-forming cardboard is approximately eight feet. The lower
    weathered claystone exerted direct uplifting pressure on the grade beam
    in this portion of the house. With 8-foot-long grade beams, 9 inches

Figure II-3. Separation of house due to differential expansion.

wide, without air space, the total uplifting pressure exerted on the
grade beam can reach as high as 96,000 lbs.   This pressure is
sufficient to cause the severe movement between the one-story portion
and the two-story portion of the house.

3.   Since the piers are only four feet in length, the soils exerted not
     only uplift pressure around the perimeter of the pier but also acted
     directly on the bottom of the piers.

The maximum possible uplift in this case is as follows:

Data: Pier diameter   = 12 in.
                Pier circumference   = 3.14 ft.
                Pier end area   = 0.785 sq. ft.

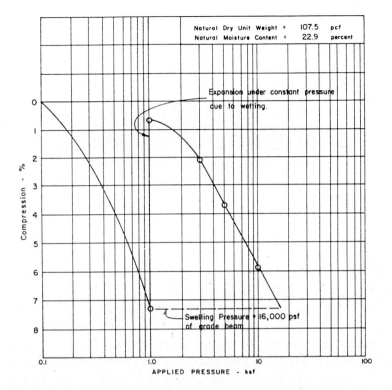

Figure II-4. Typical sample of weathered claystone obtained
from beneath grade beam.

Portion of pier in	
bedrock	= 3.0 ft.
Swelling pressure in	
bedrock	= 16,000 psf
Portion of swelling pressure responsible	
to uplift	= 16,000 x 0.15
	= 2,400 psf

then

Total uplift force	
From pier end	= 16,000 x 0.785
	= 12.5 kips

From pier wall   = 2,400 x 3 x 3.14
                                  = 22.6 kips

Total   = 35.1 kips

and

Total withholding force
(Dead load pressure)   = 15,000 x 0.785
                                  = 11.8 kips

It is obvious that the dead load pressure exerted on the pier is not sufficient to prevent uplift. The interior piers have even less dead load than the exterior piers. Consequently, the piers beneath the I-beam have lifted, causing the I-beam to move, thus disturbing the entire upper structure.

4. The partition walls in the two-story portion of the house are slab-bearing, and when the slabs heave, the walls impart direct uplifting pressure to the I-beam which disturbs the upper structure.

5. The cause of wetting of the soils beneath the foundation is from a high water table and poor drainage around the house. Initial foundation investigation indicates that the water table is near the basement floor level. Initially, consideration should have been given to the effect of wetting on the structural stability.

The cause of movement of this house is due to the swelling of the soils beneath the grade beam and the uplifting of the piers. The design and construction of the house cannot accommodate the severe uplifting of the soils.

REMEDIAL MEASURES

The following remedial measures were recommended:

1. Excavate around the basement portion of the house to expose the grade beam. Remove all mushrooms above the piers and reconstruct the air space in the same manner as in the crawl space portion of the house.

2. Free the piers from the grade beam.

3. Precise leveling should be made using the central grade beam as a reference point to relevel the entire house. It is possible to definitely establish the amount of adjustment which is required for

each individual pier to return the house to a level position. It is expected that afte this has been done, the existing cracks will be partially closed.

4. The piers, after adjustment, should be shimmed with steel plates.

5. The backfill in the basement portion of the house should be provided with deep wells approximately 3 feet in diameter around the five exposed piers so that in the future, adjustment of the piers will be possible without again removing all backfill.

6. The top of the wells should be covered with suitable material so that surface ater will not seep into the wells.

7. Readjust the I-beam to level the upper structure.

8. Remove all slab-bearing structures, such as the stairway, interior cupboards, bookcases, furnace, and so forth, and provide slip joints so that further slab movement will not affect the upper structure.

9. Check the grade beam beneath the walk-out door at the rear portion of the house to insure that the grade beam is tied in as a unit. If necessary, new grade beams should be cosntructed to span above the walk-out door.

10. Remove the rear patio slab for the entire length so that the slab will be free from the grade beam.

11. Free the basement floor slab around the perimeter of the grade beams.

12. Extend the exterior subfdrainage system from the rear side of the house to the south of the garage to intercept all possible sources of free water from entering the house.

13. Recompact the backfill in thin, moistened layers with a mechanical tamper.

14. Regrade the backfill around the house so that surface water will drain away from the house.

15. Remove all shrubs and flower beds from around the house and extend the downspouts.

This specific case went to the court and the contractor was ordered to pay the $11,000 cost of remedial construction which amounted to about 50 percent of the cost of the house. The remedial measures were completely carried out.

Shortly after the house was releveled, some of the more severe cracks started to close as shown on Figure II-5. It took more than 6 months before the structure was stabilized. Some 6 years after the remedial construction, no serious foundation movement had taken place in this house (Figure II-6); therefore, readjustment of the piers was not necessary.

Figure II-5.   Movement of the house before and after remedial correction.

Figure II-6. Condition of front of house in 1974.

Case III

DISTRESS CAUSED BY HEAVING OF
FOOTING PAD AND FLOOR SLAB

GENERAL

This case study is typical of that of underestimating the swelling potential of the soil. Individual foundation pads have heaved and severe floor heave has taken place. Because of cost, remedial measures were only partially carried out. Fifteen years have elapsed since the remedial work was completed and the buildings remain in perfect condition.

HISTORY

The buildings under investigation are in a State-operated ward for housing the severely retarded and consist of six cottages, three to the east and three to the west. The three cottages located at the western portion of the site are identified as Cherub, Aspen, and Birch. The eastern cottages are the Starlight, Crescent, and Buttercup. Each group is connected to the service buildings as indicated on Figures III-1 through III-4.

The cottages were completed in 1962. The original soil report recommended that the buildings be founded with spread footings on a combination of compacted fill and the in-place natural sandy clays, designed for a maximum soil pressure of 3,000 psf and a minimum dead load pressure of 1,500 psf. The structural design of the buildings indicates that they were founded on individual pads designed for the pressures recommended. The footings and slabs are founded partly on compacted fill and partly on natural soils. The fill was compacted to 100 percent standard Proctor density under the footings and to 95 percent standard Proctor density under the slabs.

Cracks first appeared in the building in 1963 and have continued steadily since. An investigation into the cause of cracking was made by a soils engineer in October 1966. At that time, it was recommended that drainage around the exterior of the buildings be improved.

Figure III-1. View of cottages.

DISTRESS

     In general, the extent of cracking is more severe at the eastern group
of cottages than at the western group. Typical kinds of cracking which took
place at the various buildings are as follows:
1.   Tension cracks near the top of the concrete columns,
2.   Corridors which connect the service buildings to the cottages were
     separated both horizontally and vertically.   Diagonal cracks are
     general in the brick at the junction of the corridors with the service
     buildings,
3.   Slab bearing partition walls were cracked and slightly buckled,
4.   The entrance to the service buildings had moved and diagonal cracks
     were found in the brick veneer, and
5.   A portion of the floor slabs had moved with respect to the grade beams.
     the extent of cracking in the Aspen and Cherub cottages was less severe
than in the other four cottages. Typical distresses are shown on Figures
III-5 through III-7.
     The following slab movement data was obtained:

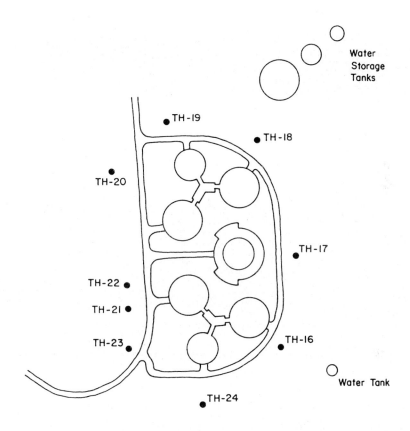

Figure III-2. Location of exploratory holes for the cottages.

Buttercup	2.3 in.	Buttercup to Service	2.1 in.
Crescent	4.4 in.	Crescent to Service	0.3 in.
Starlight	4.1 in.	Starlight to Service	0.7 in.
Cherub	2.3 in.	Cherub to Service	1.2 in.
Birch	3.1 in.	Birch to Service	0.9 in.
Aspen	2.4 in.	Aspen to Service	2.0 in.

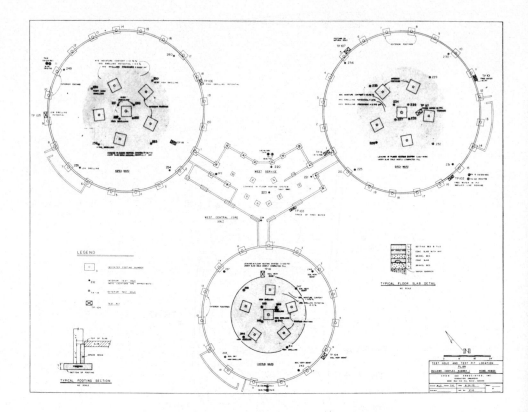

Figure III-3. Plan of test hole and test pit location for west cottages.

The above data indicates that the differential slab movement is the greatest at Crescent, while the Buttercup building shows only a 2.3 inch differential slab movement. Considering the amount of wetting of the slab in this building (Buttercup), it is likely that the slab raised more uniformly than in the other buildings. The corridors connecting the service buildings to the various cottages indicate definite movement. This movement can be associated with the different loading conditions in the cottages with respect to the corridors.

INVESTIGATION

In August 1968 investigation into the cause of cracking of the buildings, the source of moisture that entered the subsoils and the

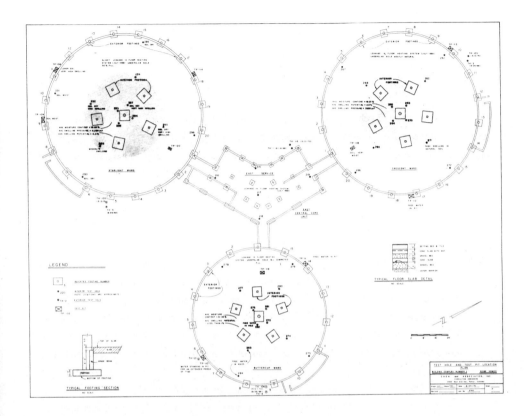

Figure III-4. Plan of test hole and test pit location for east cottages.

possibility that a neighboring water tank had an undetected leak that
provided the moisture was initiated, as well as the possible remedial
measures needed.

    In the first phase of the investigation, 24 exploratory holes were
drilled at the site, 15 of which were drilled adjacent to the cracked
buildings. The remainder of the test holes were drilled away from the
cracked buildings at locations shown on Figures III-2.

    A complete report, including recommended remedial measures, was
submitted in September 1968. For more than 1 1/2 years, no corrective
action was taken. In the meantime, the condition of all buildings continued
to deteriorate. Aspen and Cherub, which were in relatively good condition
in 1968, now showed severe movement. The movement of the various buildings

became so severe that it was necessary to evacuate the patients from all six buildings.

In April 1970 a decision was made to initiate a second phase of investigation. With the two phases of investigation, all possible factors that could influence the effectiveness of the remedial construction would be covered.

In the second phase of the investigation, the concrete slab was core drilled in twelve locations in each building and hand augered in the core hole to a depth of six feet. Undisturbed samples were obtained in each auger hole. A total of 82 holes was drilled inside the building. The location of all test holes and test pits is shown on Figures III-2 through III-4.

The investigation was directed mainly toward the following items:

1. Determination of the variation of moisture content in the soil beneath the floor slab in each building to a depth of approximately six feet.
2. Determination of the swelling potential and the swelling pressure of the soils beneath the floor slabs at various depths.
3. Determination of the swelling potential and the swelling pressure of the soils directly beneath the exterior footings at each building.
4. Determination of the possible sources of moisture which entered the buildings.
5. Determination of the water table elevation in the area.
6. Prediction of the future behavior of the soils and of the effectiveness of the proposed remedial measures.

The behavior of the soils involves many variables some of which cannot be determined with certainty. This investigation is based solely on the statistical average behavior of the soils rather than the result of a single observation or test. The evaluation of the behavior of the soils is much more complicated and difficult than for other elastic engineering materials. In this investigation, more than 150 tests on swelling characteristics, over 300 tests on moisture content, and many other tests were made so that proper conclusions could be drawn along with recommendations for remedial measures.

Subsoil conditions

Subsoil conditions at the site consist essentially of 0 to 8 feet of fill overlying soft to stiff clays. Bedrock was found at depths ranging

Figure III-5. Upper rounds – West complex– Cherub ward –
Wall braced to prevent falling in.

from 8 to 29 feet.   The characteristics of the various subsoil strata are
described as follows:

Fill--The fill consists of the on-site soils and it was often difficult to
distinguish between the fill and the natural soil.   The fill was placed
under controlled condition and the optimum moisture content ranged from 16.3
to 18.6 percent.   The actual moisture content of the in-place fill ranged
from 10.7 to 16.1 percent.

Clay--The clays at the site had fairly uniform characteristics.   The soil
could be classified as on the borderline between CL and CH with the liquid
limit ranging from 42.0 to 53.6 percent and the plasticity index ranging
from 26.8 to 32.5 percent. The stiffness of the clay varied with the

Figure III-6. Corner of Cherub ward. Interior wall pulling
away from exterior wall.

moisture content. In general, the upper soils were soft and their stiffness
increased with depth.

An X-ray diffraction analysis indicated that the total clay mineral
(not including clay-size quartz or calcite, etc.) was probably not more than
5 percent by volume of the total sample. Major minerals were quartz and
calcite, especially in the decanted fractions. Minor minerals were
montmorillonite (unusually broad 14-angstrom lines) with possibly a trace of
kaolinite and mica.

Hydrometer analysis indicated that the clay fraction (percent minus
0.002 mm) of the typical sample was less than 35 percent, and the colloid
content (percent minus 0.001 mm) less than 22 percent. The shrinkage limit
of typical clays ranged from 9.7 to 14.0 percent.

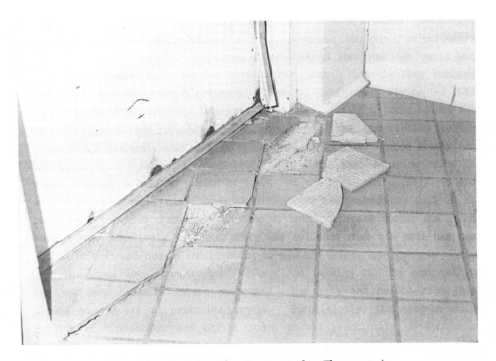

Figure III-7. Interior Aspen ward - Floor crack.

The above physical analysis of the clay soils indicated that in accordance with the established methods of classifying expansive soils, the foundation soil under the various buildings falls into the category of "highly expansive soils."

Bedrock--Bedrock consisted of claystone and sandstone. The upper portion of claystone was highly weathered. The claystone bedrock had a resemblance to stiff clay and it was difficult to distinguish between the claystone and the upper clay. The claystone bedrock had essentially the same physical characteristics as the upper clay.

Stabilized free water in the area was found at depths of 5 to 19 feet below the top of the floor level. However, at the site of the buildings, the water table was at least ten feet below the floor level.

Method of approach

After all the test data had been accumulated, the method of approach
used in solving the problem was as follows:

Swelling Potential--Swelling potential is an index that indicates the degree
of volume change of the soil after saturation. From the swell potential, it
is possible to estimate the magnitude of floor and foundation heaving. The
swelling potential of soils under each building was obtained and a curve
which graphically summarized the results was proposed by plotting the
moisture content versus swelling potential. A curve was prepared for each
building (Figure III-8).

Swelling Pressure--Swelling pressure can be defined as the pressure required
to keep the volume of the sample constant. For constant density, swelling
pressure should have a constant value. With variable moisture content and
density, the swelling pressure varies as shown on Figure III-9. Figure
III-9 is a typical graph of swelling pressure versus moisture content for
each building.

Average Moisture--The moisture content of all soil samples beneath both the
slab and the footings was obtained. An average moisture content was
determined, which provided information on the following:
1.   Average moisture content for the entire building at various depths.
2.   Average moisture content in the perimeter of the building at various
     depths, and
3.   Average moisture content at the central portion of each building at
     various depths.
     The recommended remedial measures for each building are essentially
based on the swelling pressure, swelling potential, and moisture
distribution. The conclusions are based on the statistical average of soil
behavior.

Source of moisture

No volume change will take place in the expansive soils unless there is
a change in the amount of moisture in the soil. The increase of moisture
content can be caused by various factors as follows:

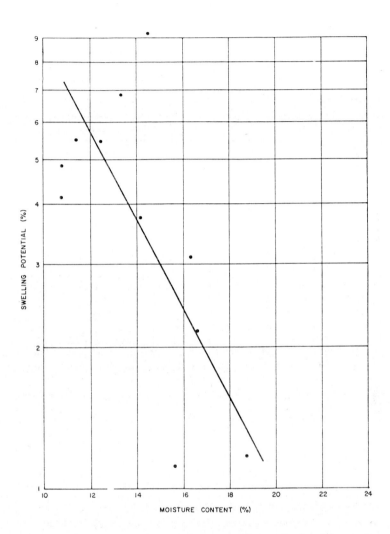

Figure III-8. Moisture and swelling potential relationship at Aspen.

1. Surface runoff including rain, melting snow, and lawn sprinkler water,
2. Leaks in the underslab heating system,
3. Leaks in the sewer system,
4. Leaks in the domestic water system, and

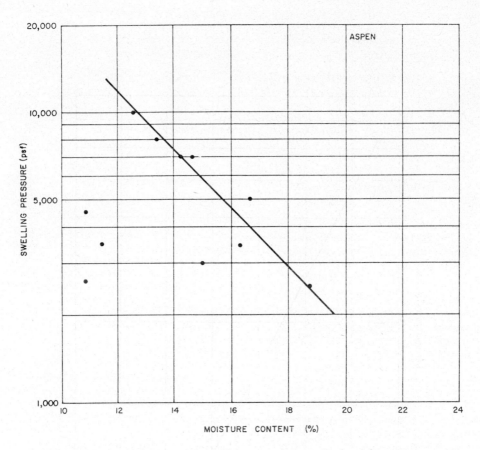

Figure III-9. Moisture and swelling pressure relationship at Aspen.

5.   Possible rising water table condition due to increase in subsurface
     water volume.

     The most difficult aspect in investigating the source of water was in
determining whether the underslab soils were wetted by the introduction of
surface water due to poor exterior drainage or by leakage of the underslab
utility system. Since each building is surrounded by grade beams three feet
deep, surface water can enter the subsoils only at a depth of at least three
feet below the top of the floor slab. However, it is most likely that
exterior surface water has entered the underslab soils through the void
space beneath the grade beams; therefore, the following conclusions can be
established:

1. If the moisture content directly beneath the concrete slabs (within 24 inches below the top of the floor slab) is high, then a leak in the underslab utility lines is suggested.

2. If the moisture content around the perimeter of the building at a depth of more than three feet below the floor slab is high, then the migration of exterior surface runoff into the underslab soils is suggested.

3. If the moisture content around the perimeter of the building and the moisture content of the building's interior are both low, then no introduction of surface runoff or leakage in the utility lines is suggested.

4. If the moisture content around the perimeter of the building and the moisture content of the building's interior are both high, then both the introduction of surface runoff and leakage in the utility lines are suggested.

The average moisture content at various depths for each building will give a clear indication as to the source of moisture that has entered into the buildings.

TREATMENT

Based on the above reasoning, the problem that existed in each building can be established and remedial measures prescribed.

Treatment at Birch

Birch is the east building of the west complex. Foundation soils at this building site consist essentially of controlled compacted fill. A study of the original soil report indicates that at the north side of this building there is approximately 1/2 foot of cut and at the south side there is approximately 7 feet of fill.

Three test pits were opened at the exterior of the building, adjacent to the grade beam. In Test Pits 101 and 102, water was flowing from underneath the slab. Further testing indicated that the underslab heating system had leaked and the sewerline had broken. This resulted in the flooding of the underslab soils.

Most of the compacted fill soil beneath the footings possessed only low swell potential. The possibility of foundation movement is relatively slim. Moisture contents for the entire building at various depths are relatively uniform with the lowest moisture content 15.8 percent and the highest moisture content 22.6 percent. Further slab movement should not exceed 1/2 inch.

The following remedial measures are recommended for this building:

1. The sewerline which runs under the building and branches into the various bathrooms should be exposed and carefully checked for leakage.

2. Underpinning of the exterior or interior footings will not be necessary. Remedial measures to the footings are not recommended.

3. The perimeter of the floor slabs should be saw cut to insure that the slab is separated from the grade beams and that there will be free movement of the slab with respect to the grade beams.

4. The slab-bearing partition walls in this building should be reconstructed in such a manner that slab movement will not affect the stability of the structure. A vertical slip joint should also be provided where the partition walls connect with the exterior walls or columns.

Treatment at Aspen

Aspen is the north building of the west complex. The foundation soils at this building consist entirely of the natural soils. The amount of cut in the site grading ranged from 5 to 20 feet.

The swelling potential of the soils beneath the exterior footings is high with a percent of swell of 5.5 percent and swelling pressure of 10,000 psf. The average moisture content beneath the interior footings was 14.7 percent. This corresponds to an average swelling potential of 3.2 percent, (Figure III-8) and an average swelling pressure of 6,200 psf (Figure III-9). At a depth of six feet below the top of the slab, the average moisture content decreased to 11.6 percent. This corresponds to an average swelling potential of 6.5 percent and average swelling pressure of 14,000 psf.

The moisture distribution indicates that the lower soils are in a very dry state, and if the soils become excessively wetted, swelling will take place.

In 1968, a mechanical engineer found that there was only slight leakage in the underslab hot water heating system by conducting pressure tests. Another pressure test was conducted in April 1970, and the pressure dropped from 130 to 27 psi in twenty minutes. It was obvious that in the preceding months more leakage had developed in the underslab heating system which accounted for the severe movement in this building.

The soils directly beneath the floor slab had an average moisture content of 13.8 percent. This is low compared with underslab moisture content of the other buildings. The movement of the floor slab in this building had only begun and future severe floor movement will take place even though the underslab heating system is entirely disconnected. The existing pockets of high moisture content soils caused by leakage of the heating system will migrate to the drier phase of soils and cause damage. This cannot be prevented unless all problem soils beneath the slab are removed.

The following remedial measures were recommended for this building:

1. Underpin the exterior footings with piers drilled into bedrock. The piers should be designed for a maximum end pressure of 30,000 psf and a skin friction of 3,000 psf for that portion of the pier in bedrock. The piers should also be designed for a minimum dead load pressure of 20,000 psf. The piers can be drilled in a slanted position, or two piers can be drilled under each column with a grade beam spanning over the two piers.

2. The interior footings should also be underpinned; however, it is difficult to underpin the interior footings without demolishing the entire building. Therefore, it was recommended that the interior footings be decreased in area by cutting off the concrete pad and thus increasing the unit dead load pressure. It was estimated that the dead load pressure on each pad could be increased to over 6,000 psf by reducing the area of the concrete pad.

3. The floor slabs in this building should be entirely removed and the soils beneath the slab removed for a depth of three feet. These soils should be discarded and replaced with nonexpansive, impervious, granular soils compacted to at least 90 percent standard Proctor density at optimum moisture content.

If for some reason the underslab soils cannot be removed, there is every possibility that the floor slab will raise as much as three inches above the present level. Slip joints in the partition walls will prevent the disturbance of the upper structure, but unsightly cracks in the partition walls and the floor slab will take place.

Treatment at Cherub

Cherub is the west building of the west complex. The foundation soils beneath this building are mostly fill. At the north side, there is one foot of cut and at the south side there is about seven feet of fill. Bedrock is shallow at the west side of the building.

The average moisture content of the soils beneath the exterior footings was about 22.8 percent which corresponds to a swelling potential of less than one percent. The possibility of foundation movement was rather remote. Most of the interior footings are placed on natural soils. The average moisture content of the soils beneath the interior footings was about 14.5 percent. This corresponds to a swelling potential of 5.0 percent. Since the moisture content was low, the chance of increase in moisture content beneath the interior footings is high and there is a strong possibility that the footing foundation will have future movement.

In 1968, a mechanical engineer again conducted pressure tests on the underslab hot water heating system. The tests indicated that there was no leakage in the system. On January 28, 1970, similar tests were made in this building which indicated that the pressure dropped from 90 to 76 psi in fifteen minutes and from 100 to 52 psi in 75 minutes. This indicates that there is leakage in the underslab heating system.

The moisture content distribution analysis indicates that the moisture content near footing level was higher than for the soils directly beneath the floor; also, the exterior moisture content was generally high, about 4 percent higher than the interior moisture content. This definitely indicated that most of the wetting of this building had been caused by the migration of surface water into the foundation soils. The leakage of the underslab heating system had taken place only recently and the effect of the leakage had not been reflected in the moisture content of the soils.

The soils directly beneath the floor slabs had an average moisture content of 12.1 percent which corresponds to a swelling potential of more than 8 percent. The movement of the floor slab of this building has only begun and severe floor movement will occur even though the underslab heating system is entirely disconnected. The existing local high moisture content in the soil will migrate to the drier soil and cause floor damage.

The following remedial measures are recommended for this building:

1.   It is not necessary to underpin the exterior footings. The moisture content around the perimeter of the building is high and further swelling of the soils beneath the footings is unlikely.

2   There is a strong possibility that the interior footings will have movement. The interior footings should be decreased in area by cutting off the concrete pad and thus increasing the unit dead load pressure, and

3.   The floor slabs in this building should be entirely removed and the soils beneath the slabs removed to a depth of three feet. These removed soils should be replaced with nonexpansive, granular soils compacted to at least 90 percent standard Proctor density at optimum moisture content. If this is done, then the possibility of further floor movement will be remote.

Treatment at Buttercup

Buttercup is the east building of the east complex. The foundation soils at this building site consist of two to five feet of controlled compacted fill. It was suspected that all footings, both interior and exterior, were founded on structural fill.

The average moisture content of the soils directly beneath the footings was 25.6 percent. This corresponds to a swelling pressure of 2,000 psf and a swelling potential of less than one percent. At a lower depth, the natural soils were generally dry with the moisture content ranging from 18 to 20 percent and the swelling pressure reaching as high as 9,000 psf.

Judging from the moisture condition of the soils beneath the exterior footings, it was not necessary to underpin the exterior footings. The possibility of footing foundation movement was relatively remote.

The average moisture content beneath the interior footings was 22.1 percent. This corresponds to an average swelling potential of less than one percent. The soils beneath the exterior footings had relatively uniform moisture content with a minimum moisture content of 20.1 percent and maximum moisture content of 26.9 percent. The possibility of foundation movement for the interior footings was also remote. It was, therefore, not necessary to underpin or make remedial construction on the interior footings.

In 1968, a pressure test was made on the underslab heating system, the results of which showed that the pressure dropped from 32 to 0 psi in 60 seconds. This definitely indicates that there is a large leak in the underslab heating system. Free water was found not only in the exterior test pits but also in three test holes inside the building. The amount of water trapped in the underslab soils must be near saturation which accounts for the steady flow of water from the underslab soils in Test Pit 110.

Also, sewer tests that were conducted indicate that exterior test pits had filled during the tests. It was concluded that there are very definitely leaks in the sewerlines of the east complex.

The moisture content of the soils at the perimeter of the building was only slightly higher than the moisture content at the central portion of the building. This indicates that the amount of water that seeped into the soils from the exterior of the building was relatively low. Most of the water present in the underslab soils is derived from the leakage of the underslab heating system and possibly from leaking sewerlines.

The possible behavior of the floor slabs at this building can be evaluated by studying the moisture distribution at various depths. The following facts were noticed:

1. The soils directly beneath the slabs and an average moisture content of 21.0 percent. This corresponds to an average swelling potential of 1.2 percent which is considered to be low.

2. There was no large difference between the moisture content in the center portion and the moisture content around the perimeter of the building (19.9 versus 21.2 percent); therefore, the migration of moisture from interior to exterior is unlikely to take place.

3. Due to the excessive leakage of the underslab heating system, a large portion of the soils had reached a state of saturation.

It is not necessary to replace the underslab soils and the chance and magnitude of future slab movement is low.

All backfill around this building should be removed to expose the foundation system. After the backfill has been removed, the excavation should remain open for a period of at least two weeks to insure that all water trapped under the floor slabs can be effectively drained.

Treatment at Starlight

Starlight is the east building of the east complex. The foundation soils at this site consist of both cut and fill. At the north side, there is two feet of cut and at the south side there is four feet of fill. It is likely that all footings, both exterior and interior, are founded on the natural soils.

The swelling potential beneath the footings was erratic, ranging from 0.3 to 3.4 percent, with swelling pressure ranging from 2,500 to 12,000 psf. Since the footings are only lightly loaded, further movement of the footings is possible. A study of the average moisture content at the perimeter of the slab indicated that the average moisture content was 21.1 percent, which corresponds to a swelling potential of 1.5 percent. The potential movement of the exterior footings is at least 50 percent.

The average moisture content beneath the interior footings was 15.0 percent. This corresponds to an average swelling potential of 4.5 percent and an average swelling pressure of 6,000 psf. At the lower depth, approximately six feet below the top of the slab, the average moisture content decreased to 14.1 percent. This corresponds to an average swelling potential of 6.0 percent and an average swelling pressure of 15,000 psf. It was obvious that the potential movement of the interior footings is high and in fact, at a depth of six feet below the top of the floor slab, some of the soils had a swelling potential as high as 9 percent and a swelling pressure as high as 30,000 psf.

Again in 1968, the underslab heating system was checked and only slight leakage was found. During this test, the pressure only dropped from 34 to 33.5 psi.

None of the test pits nor test holes, either inside or outside of the building, showed signs of the presence of free water. This indicated that leakage of the utility lines was not the main problem at this building. The perimeter moisture content at a depth of three feet was 21.1 percent and the moisture content at the central portion of the building at the same depth was 15.0 percent. This definitely indicated that the source of moisture that entered this building was from surface runoff. Poor drainage around the building was directly responsible for the wetting of the foundation soils.

The soils directly beneath the floor slab had an average moisture content of 16.9 percent, corresponding to an average swelling potential of 3.5 percent. Moisture distribution in the underslab soils was extremely erratic, with the highest moisture content being 22.7 percent and the lowest 10.6 percent. The potential for further floor movement at this building is great. Moisture distribution was erratic, and even though the source of moisture that entered the building had been cut off by improving the drainage, there is still the chance of moisture migration from the wet area to the dry area, and this can cause damage.

The remedial measures given under "Treatment at Aspen" can be applied in their entirety here. Since the main source of water which entered the building is from the surface runoff, special attention should be directed to improving the surface drainage condition.

Treatment at Crescent

Crescent is the north building of the east complex. The foundation soils at this site consist of part fill and part cut. At the west side there is three feet of cut and at the east side there is 1 1/2 feet of fill. It is likely that all footings at this building are placed on natural soils.

The condition of soils beneath the exterior footings is represented by Test Pits 112 and 113. Free water was found in Test Pit 112. The water seeped into the pit from the underslab gravel and drained away in several days. Soils at footing level in Test Pit 112 possessed only low swelling potential.

In Test Pit 113, no free water was found. The swelling potential at footing level was low but at lower depths, the swelling potential reached as high as 2.4 percent with a swelling pressure of 6,000 psf.

The average perimeter moisture content was on the order of 21.2 percent. This corresponds to an average swelling potential of 1.8 percent and a swelling pressure of 3,000 psf. From the study of the soil behavior beneath the exterior footings, the possibility of foundation movement is remote. It is not necessary to underpin the footings.

Most of the interior footings are placed on natural soils. The average moisture content of the soils beneath the interior footings was 19.2 percent. This corresponds to an average swelling potential of 2.3 percent and an average swelling pressure of 3,500 psf. The moisture content decreased with depth. At a depth of six feet below the top of the floor slab, some of the natural soils possessed high swelling potential, as much as 8.3 percent. There is some chance of movement of the interior footings. Again in 1968, definite leakage was found in the underslab heating system. Tests indicated that the pressure dropped from 35 to 22 psi in five minutes.

A study of the moisture content of the underslab soils indicated that the moisture content directly beneath the floor slab was high and the perimeter moisture content was even higher than the central moisture content. This indicated that the underslab soils were wetted by leakage of the utility lines as well as by surface runoff.

The soils directly beneath the floor slab had an average moisture content of 20.5 percent. This corresponds to an average swelling potential of 1.9 percent and an average swelling pressure of 3,000 psf. The moisture content was relatively uniform within four feet below the surface of the slab. With proper drainage, the chance of further floor movement is not great.

The remedial measures for this building are essentially the same as those recommended for Birch. For the floor slabs, the recommended procedure is a choice between removal of the soils beneath the slab as recommended for Aspen or saw-cutting the floor slabs and improving the partition walls as recommended for Birch.

Drainage improvement

In addition to the remedial measures given for each building, the following general treatment for improving the exterior drainage was recommended:

1. Remove all backfill to expose the foundation system. After removal a careful check should be made of the void space beneath the grade beams to assure that it is properly formed. The excavation should remain open for a period of at least one week so that all water which is trapped under the floor slab can be drained out.

2. A concrete walk should be provided around the exterior of each cottage. The walk should be at least eight feet wide with sufficient slope to allow free drainage of water away from the cottages. The walks should not be tied into the grade beams. An expansion joint should be provided between the walk and the grade beam.

3. The ground surface surrounding the cottages should be sloped to drain away from the cottages. A slope of ten inches for the first ten feet is recommended. If this slope is unattainable, it may be necessary to install swales at various locations outside of the cottages to lower the ground surface and allow free drainage.

4. The lawn sprinkler heads should be located at least ten feet from the foundation walls of the buildings. Spray from the sprinkler heads should not be directed towards the buildings.

5. The roof downspouts are not efficient. It is entirely possible that during heavy storms, water could collect, overflow, and not drain away through the downspouts. Improvement is necessary in this area.

6. Drainage is notably poor in front of the service buildings. These areas should be paved with concrete if positive surface drainage is not possible.

REMEDIAL CONSTRUCTION

Remedial construction started in September 1971. All recommendations were carried out except the underpinning of the exterior footings. Budget limitations did not allow the full remedial construction. The remedial construction, at a cost of a half million dollars, was completed in July, 1972. Figures III-10 through III-15 show certain construction procedures.

Considering the extent of original damage, the remedial construction has proven successful. It is unfortunate that the severe swelling potential of the subsoil had not been recognized in the design stage. Had the buildings been founded on piers rather than on individual pads and had the underslab heating system been eliminated, most of the damage could have been avoided.

Figure III-10.   Installation of subdrains around the perimeter of
the building.

Figure III-11. Checking the void forming material beneath the grade beams.

Figure III-12.   Repair of leaking plumbing.

Figure III-13. Drainage around exterior of the building has been
improved. Note catch basins in the lawn area.

Figure III-15. Concrete apron placed around the building with properly constructed mastic joints.

Figure III-14. Severely cracked brick wall, patched and repaired. Note: Further cracking has not occurred.

DISTRESS CAUSED BY HEAVING OF CONTINUOUS FOOTINGS

GENERAL

This case study is typical of what happens when continuous footings are placed on expansive soil without considering uplift forces. In this instance the study involves a house that was constructed without benefit of adequate design. Structural strength in the foundation walls was lacking; therefore, it was not possible to underpin the building. Use of post-tensioned steel or pouring of a new foundation wall appears to be the only possible remedial construction. By so doing, the house will be tied together as a box and will be able to withstand further differential movement.

HISTORY

The house is located in Broomfield, Colorado, a small community north of Denver, Colorado. This area is well known for its swelling soil problem. The house was constructed in 1960. Neither a soil investigation record nor structural design drawings were available. The house has a full basement and attached garage.

Test pits were excavated to examine the foundation system of the house. This revealed that the house is founded with continuous spread footings on the natural soils. The footings are 20 inches wide and 8 inches in depth. To conform with the natural ground contour, the basement foundation wall is stepped down from full basement height at the south end to only 24 inches at the north end. The space between the concrete wall and brick course is filled with cinder block.

No reinforcement was found in the concrete foundation wall or in the footings. Such foundation design creates a structural weakness in the middle portion of the basement wall. Without reinforcement in the concrete, slight foundation movement will result in severe cracking. Also, at the north end of the basement, there is a structural discontinuity at the walk-out entrance, where the grade beam is disrupted.

None of the slab-bearing partition walls have slip joints either at the top or bottom. Any slab movement will cause the partition wall to push against the upper floor joists and cause movement in the upper levels.

Exterior

At the exterior of the house, severe cracks were found, both on the west and east sides in the brick walls above the basement. Most of the cracks appeared directly below and above windows in the middle portion of the basement section of the house. Cracks opened (Figure IV-1), as much as three-quarters of an inch, and extended to the concrete beneath the cinder block. At the north side of the building, severe cracks were also found below the windows. No severe cracks were found on the south side of the building.

Basement interior

The basement floor slab had been removed, exposing a portion of the footings. Prior to removal, the basement floor slab was badly cracked. Cracks were found in the footings as well as in the foundation walls. A trench was opened around the basement portion of the house in an attempt to install drain tile around the basement wall, leading to a sump. A 4-inch gravel layer was placed beneath the basement slab.

Upper floor

The amount of cracking in the upper floor is also severe as shown on Figure IV-2. Cracks were evident above windows, at the east side of the house, and a few doors were jammed in the upper level.

Exterior drainage

Exterior drainage conditions of the house are poor, particularly at the north side of the house where the walk-out basement door is located. Natural drainage is from west to east, and there is a strong tendency for surface water to enter the foundation soils from the west side.

SUBSOIL CONDITIONS

The subsoil conditions at the site consist essentially of slightly porous sandy clays at the south end, and highly weathered, moist claystone at the north end. Undisturbed hand drive samples were taken from the test

Figure IV-1. Cracks below window.

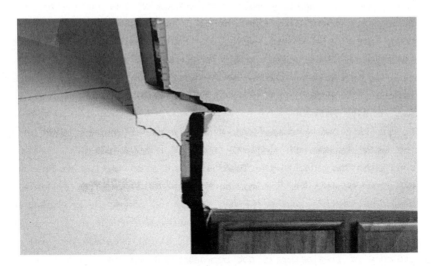

Figure IV-2. Severe cracking above closet

pits. Swell-consolidation tests, performed on the undisturbed hand drive samples, indicated that the upper sandy clays at the south side of the building possessed only low swelling potential; while the weathered claystone at the north side of the building possessed moderate to high swelling potential. The soils, at present, are in a very moist condition.

It is apparent that the soils were in a much drier condition when the house was constructed, and the swelling potential of the claystone should be much higher than the tests indicate.

No free water was found in the test pits which had a depth of five feet beneath the bottom of the footings.

CAUSE OF MOVEMENT

The cause of movement of the house is due to a combination of uplifting movement of the foundation soils and poor structural design as summarized below:

1. There is a difference in depth of the concrete foundation walls, and also the foundation walls are not reinforced. The structure of the house cannot withstand even slight differential movement. This is substantiated by the fact that the foundation wall, as well as the foundation, have both cracked severely.

2. The amount of swelling of the claystone at the north end of the building is about five times as much as the amount of swell of the sandy clays at the south end of the building under the same moisture conditions. Consequently, the foundation at the north end of the building has lifted. Measurement confirms that there is a difference in elevation between the north and south ends of the building by as much as five inches.

3. At one time, water seeped into the basement at the west side through the seams between the concrete and cinder block wall. Such wetting conditions have caused the foundation soils to swell. No free water was found beneath the footings at the time of inspection. Judging from the high moisture content of the foundation soils, it is evident that the entire area has been under severe wetting conditions. The swelling of the weathered claystone requires only slight moisture increase. The presence of free water is not necessarily the cause of these swelling conditions.

REMEDIAL MEASURES

Since the cause of movement is due to both the structural weakness of the building and the swelling of the claystone beneath the footings, the following remedial measures are recommended:

1. A new foundation wall should be poured around the interior of the basement. The new grade beams should be reinforced with two 5/8-inch bars, top and bottom, and should be tied in with the existing foundation wall in the manner shown in Figure IV-3. The new foundation wall should have a depth of approximately the full height of the basement. This wall will tie the basement together structurally and eliminate the existing structural weakness.

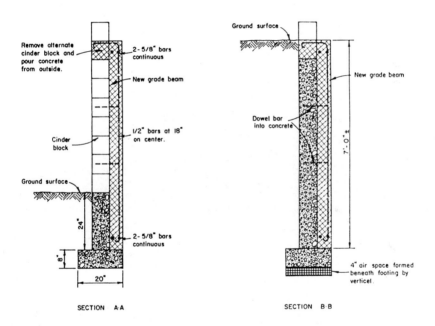

Figure IV-3. Sketch of new grade beam for basement.

2. The soils should be removed from beneath the footings at approximately 10-foot intervals, as shown in Figure IV-4. In so doing, the weight of the building will be concentrated at isolated locations and dead load pressure increased. It is estimated that the amount of dead load pressure exerted on the exterior footings is about 600 psf. Such dead load pressure is insufficient to prevent the uplifting of the claystone beneath the footings. With voids formed beneath the footings, the dead

load pressure will be increased to about 3,000 psf, eliminating the uplifting problem. Air space beneath the footings should be formed by the use of void forming material.

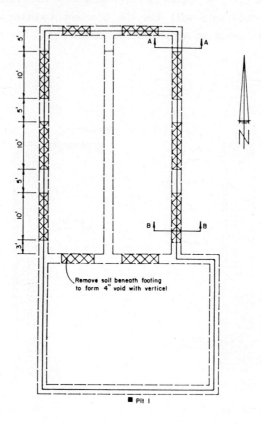

Figure IV-4. Removal of soils beneath the footings.

3.   An alternative remedial construction method is installation of post-tensioned steel cables in both the outside and inside of the foundation walls in the manner indicated in Figure IV-5. The purpose of the post-tensioned cables is to tie the entire structure together to prevent unequal movement.

4.   The drainage around the building should be improved so that water will drain away from the building. All shrubs and flower beds adjacent to

the building should be removed and all roof downspouts should extend well beyond the limit of all backfill.

5. The installation of a subsurface drainage system will not improve the present situation. However, the use of a subsurface drainage system will keep the water table below basement level. The use of a gravel layer beneath the slab will not prevent cracking of the slab, but will break any capillary water rise.

If the remedial measures presented above are performed, it is believed that movement of the house will stop, but it will take a period of at least six months before equilibrium can be established. Interior decorating and repairs should not begin until equilibrium has been established. Elevation pins should be established around the house and records kept as to the amount of movement.

Remedial construction was started shortly after the house was investigated. The post-tensioned cable system was used to strengthen the foundation (Figure IV-6). The residence is in good condition after the remedial construction.

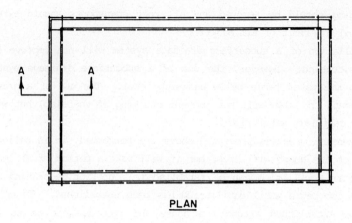

PLAN

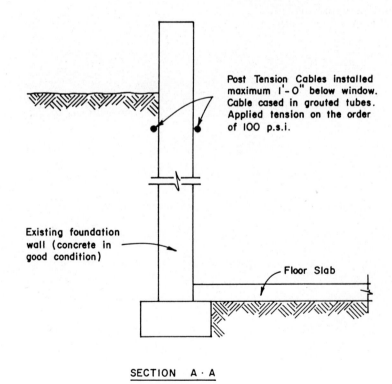

Post Tension Cables installed
maximum 1'-0" below window.
Cable cased in grouted tubes.
Applied tension on the order
of 100 p.s.i.

Existing foundation
wall (concrete in
good condition)

Floor Slab

SECTION   A·A

Figure IV-5.   Remedial measures using post-tensioned steel cables.

Figure IV-6. Post-tensioning the foundation wall.

DISTRESS CAUSED BY
RISE OF WATER TABLE

GENERAL

This case study involves 39 two-story townhouses founded on a drilled pier system, that had foundation movement. The movement of the drilled pier system is essentially caused by a rise of ground water. The following knowledge was gained from this study:

1.   In areas where there is a strong possibility of rise of ground water, the use of a drilled pier system should be carefully considered.

2.   Heaving of the floor slab can transmit high swelling pressure to the grade beams.

3.   Chemical treatment of the underslab soil, in this case, is ineffective.

HISTORY

Thirty-nine townhouses are under investigation for foundation movement. The townhouses are located in southeast Denver, Colorado. There are four to seven units in each building. A total of 251 units and a clubhouse were studied (Figure V-1).

Initial subsoil investigation for the site was made in 1965. The report resulting from that investigation recommended that the buildings be founded with piers drilled into bedrock designed for a maximum end pressure of 20,000 psf and a minimum dead load pressure of 10,000 psf.

Construction of the building complex was started in September, 1965. Shortly after the completion of the various units, movement of the buildings began, and in some buildings severe movement of both foundation walls and floor slabs was noticed. In March, 1970, a consulting soil engineer was engaged to make a preliminary investigation into the cause of cracking and movement in the various units. A report outlining remedial construction was prepared but no corrective action was taken.

The following typical distress was observed in most buildings:

Foundation walls

Cracks were found in the brick course in a diagonal pattern from the top of the window or door on the ground level to the bottom of the window in the upper level.   Most of the cracks had been patched and some had reopened. The width of the cracks ranged from hairline thickness to as much as one inch as shown on Figures V-2 and V-3.   These cracks definitely indicated pier movement.   In addition, both vertical and diagonal cracks were found in the basement foundation walls.

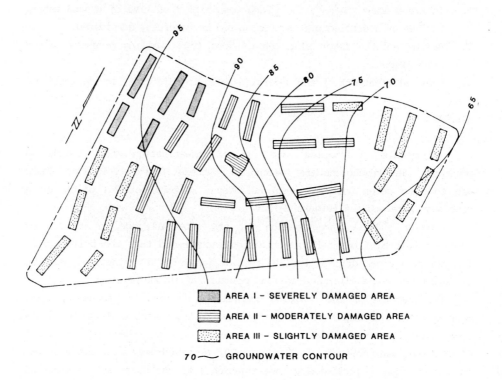

AREA I – SEVERELY DAMAGED AREA

AREA II – MODERATELY DAMAGED AREA

AREA III – SLIGHTLY DAMAGED AREA

70 GROUNDWATER CONTOUR

Figure V-1. Building location plan and ground-water contour.

Interior floor slabs

Most of the basement areas were finished.   Concrete floor slabs were covered with tile or carpeting. The slabs had heaved in many buildings and the cracks generally followed a pattern parallel to the foundation walls.

Figure V-2. Typical cracking of the townhouse exterior walls.

This is a typical slab crack where separation between the slab and foundation wall was not properly constructed. The slabs bind on the foundation walls and are not free to accept vertical movement. Consequently, cracks appeared parallel to the foundation walls. Such

Figure V-3.   Typical cracking of townhouse exterior walls indicating pier uplift.

movement not only caused the slab cracking, but also exerted uplift pressure on the foundation walls.

Partition walls

All buildings have units with partitions and in most cases slip joints were not provided in the slab bearing partition walls. Consequently, when the slabs moved upward, they exerted uplifting pressure on the upper structure. As a result, cracks developed in the upper stories, jamming the doors and distorting the floor system.

In those townhouses which had basement areas that were not finished, slab movement does not contribute to the distress of the upper stories; however, the staircase walls in all cases rest directly on the basement floor. Therefore, slab movement can transmit movement to the upper stories through the staircase partition walls.

Slab treatment

Between February and April of 1966, in several buildings mainly at the western portion of the site, the floor slab was removed and the soil beneath the slab injected with stabilizing chemicals in an effort to eliminate the swelling potential of the lower soils. After the new slab had been placed, movement of the floor slab was not checked and in most cases, slabs again cracked. The results of the swell tests indicated that the swelling pressure of the underslab soils ranged from 500 to 2,500 psf and the moisture contents at various depths were fairly uniform. Tests indicated that the application of chemicals on the underslab soil did not have a pronounced effect in reducing the swelling potential.

Aprons

Concrete aprons constructed at the rear of the buildings were in most cases cracked. Gaps were found between the slabs and the buildings, indicating separation. This can be caused by either the uplifting of the building relative to the slab or the settlement of the slab due to inadequate backfill relative to the building. The aprons were mud jacked in several buildings in an attempt to correct the condition.

In general, the cracks that developed in this apartment complex typify the cracks found in buildings founded on expansive soils. There is no doubt

that the movement of the buildings can be attributed to a swelling soil
problem.

SUBSOIL CONDITIONS

Subsoil conditions at the site consist of 7 to 22 feet of stiff to
medium stiff clays overlying claystone bedrock. The clays have an average
moisture content of 20.3 percent with the highest moisture content of 23.8
percent and the lowest of only 11.6 percent. This moisture content is
considerably higher than the moisture content reported by the testing
laboratories in August of 1965. At that time, the average moisture content
of the upper soil was only 14.6 percent. The swelling potential of the
upper clays ranges from 0.5 to 3 percent with the swelling pressure ranging
from 0 to 5,000 psf.

The lower bedrock consists basically of claystone and weathered
claystone. Some sandstone lenses were found in the claystone. A study of
moisture content of the claystone bedrock also indicates an increase in the
last few years. Since the moisture content of the claystone bedrock is
affected by the pattern of seams and fissures of bedrock, the increase of
moisture content is not as obvious as the increase of moisture content of
the upper clays. The swelling pressure of the lower claystone ranges from
5,000 to 25,000 psf. The average swelling pressure is about 10,000 psf.

It is concluded from the laboratory testing that the upper clays
possess only low swelling potential and the lower claystone possesses high
swelling potential. Most of the foundation wall movement is the direct
result of the swelling of the lower claystone; however, the swelling of the
upper clays is sufficient to cause the floor slabs to heave and crack.

WATER TABLE

When the soil and foundation investigation was made in August, 1965, no
free water was found in any of the twenty-one exploratory holes. Some of the
test holes were more than twenty feet deep. In July, 1970 when this
investigation was made, free water was found in almost every hole. Figure
V-1 indicates an approximate contour of equal elevation to water table.

The following conditions were observed:
1.   The water table elevation is high at the western portion of the site
      and low at the eastern portion.   A difference in the water table

elevation of 30 feet was observed between the northwest corner and southeast corner of the property.

2. The water table contour follows fairly well with the bedrock contour. Water was found at the top or immediately below the surface of bedrock.

3. The water table follows fairly well with the ground surface contour. The ground surface was high in the west and low in the east.

After carefully studying the contour of equal elevation to the water table and the general water table conditions in the area, it is concluded that in the last five years, after area development, there was a definite change of water table conditions. The general rise of the water table is essentially caused by a perched water table in the developed residential area. Bedrock in this area is shallow and composed essentially of claystone which is relatively impervious. Free water flows mostly on top of bedrock and also flows in the fissures and seams of the bedrock.

Surface water does not necessarily penetrate directly from the ground surface into the underlying bedrock but will flow from the high point to the low point along the surface of bedrock.

The rise of water table definitely has a bearing on the foundation movement of this apartment complex. As seen on Figure V-1, the buildings suffering the most severe damage are located in high water table elevation areas. Where the water table elevation is low, relatively minor damage to the building was experienced.

CAUSE OF MOVEMENT

In reviewing the foundation design, the grade beams and pier system and the results of this investigation, the following were derived:

Foundation design

The foundation design criteria calls for an end-bearing pressure for piers of 20,000 psf and a minimum dead-load pressure of 10,000 psf. The piers should penetrate the shale a minimum depth of five feet. The side of the pier excavated into the shale should be roughened to provide resistance to uplift. The drill logs provided by the driller indicated that the required pier penetration had been fulfilled in most cases. The load carrying capacity of the piers was reviewed, including the dead-load pressure requirement, and were found to be in accordance with the recommendations, except for the piers supporting interior columns and beams.

In most cases, the swelling pressure of the lower bedrock is about 10,000 psf with a few cases reaching as high as 25,000 psf. The recommended dead-load pressure of 10,000 psf is low, but considering the subsoils and the water table conditions at the time the subsoil investigation was made, the design recommendations are in a range of sound engineering practice.

Foundation construction

From the information obtained by excavating from the inside and from the outside of the building exposing the grade beam system, the construction was in accordance with the structural foundation design. In several places, mushrooms were found on top of the piers and the void forming material beneath the grade beams immediately adjacent to the piers was absent. Such defects decrease the dead-load pressure exerted on the piers; however, the effect is not sufficient to cause the present distress of the various buildings.

Slab construction

The major problem is in the area of slab construction. In principle, free floating slabs should be provided. There should be positive expansion joints between the slabs and the grade beams to allow free slab movement. At present, the joint between the slabs and the grade beams is not effective in most cases. Slab movement due to expansive soils has transmitted uplifting pressure from the slabs to the foundation walls. The swelling pressure of the underslab soil is about 2,000 psf.

In extreme cases, the pressure transmitted from the slab to the foundation wall can reach 2,000 pounds per linear foot. With the piers spaced on 11-foot centers, each pier can be subject to about 22 kips of uplifting pressure.

Pier uplifting

The most important reason for the movement of the foundation walls is the uplifting of the piers. In the original foundation design, it was assumed that the lower bedrock would not become wetted and the skin friction between the bedrock and the piers would provide an additional factor of safety against pier uplifting. With the rise of the water table, the entire lower bedrock became wetted; consequently, the skin friction value dissipated and rather than holding the piers, the swelling of the bedrock actually lifted the piers.   As an example, pier No. 8 in Building 8 was checked by calculation as follows:

Actual dead load pressure on the pier	=	20.6 kips
Pier diameter	=	10 inches
Average swelling pressure	=	10,000 psf
Pier penetration into bedrock	=	7 feet
Uplifting pressure due to saturation		
of bedrock - 7 x 2.62 x 10,000 x 0.15	=	27.5 kips
Unbalanced pressure		
27.5 - 20.6	=	7.1 kips

For interior piers where the actual dead-load pressure exerted on the pier ranges from 2.6 to 20.9 kips, the unbalanced pressure is in the range of 10 to 30 kips.

It is concluded that the wetting of the bedrock plus the pressure transmitted from slab heaving to the grade beam had lifted the piers.

In one test pit, the entire length of the pier was exposed. A gap of approximately three inches between the soil and the bottom of the pier was found. This verified that the pier actually pulled out of the ground and would have cracked had it not been reinforced.

SOURCE OF MOISTURE

The rise of ground water in this development is mainly derived from surface water. The source of surface water is from the following:

Precipitation

Precipitation, either from rain or from melting snow, can contribute to the rise of water table. Before the development was constructed, most of the precipitation drained from the site as surface runoff. Only a portion of the precipitation penetrated the ground.

With building construction, in isolated areas, precipitation will penetrate the soils through the loose backfill around each building and due to poor drainage conditions, tends to accumulate instead of running off the site. This constitutes an important factor toward the rise of ground water.

Lawn irrigation

After a development is completed, lawn irrigation in the area will generally create perched water table conditions. It is conceivable that a large amount of lawn irrigation water will travel through the upper soils and become trapped at the surface of the bedrock. This practice is not necessarily limited to the property investigated but pertains to the entire general area. High water table conditions prevailed in several areas in the general vicinity after the development was completed.

Pipe leakage

Leakage of the water lines, resulting from corroded service saddles, was found in front of Buildings 80, 83 and 84. It is not known how long the leakage had taken place, but monthly water consumption from January to October of 1969 was 24.2 million gallons. During this same period in 1970, the consumption was 28.3 million gallons, which shows an increase of four million gallons over a period of ten months. This increase in consumption can be partially explained by pipe leakage. These 4.1 million gallons of water will eventually flow on the surface of the bedrock and cause a general rise of the water table.

EVALUATION OF BUILDING CONDITIONS

For discussion purposes, the townhouse complex was divided into three areas (Figure V-I), and the evaluations are as follows:

Area I

In this area, severe foundation movement has taken place. The foundation walls as well as the floor slabs have cracked and heaved. Most of the floor slabs in this area have been replaced and the soils beneath the floor slabs have been treated with special chemicals.

The water table elevation is high in this area and the bedrock has been wetted excessively. Foundation movement in this area is caused by the uplifting of the piers.

Area II

In this area, relatively minor foundation movement has taken place.
Some of the floor slabs have been replaced and treatment of the underslab
soils was made in several buildings.

The buildings in this area have not suffered severe foundation movement
which is probably due to the following:

1.   The water table elevation is relatively deep and the bottoms of the
     piers are above the present water table. The piers still maintain an
     anchorage effect and expansive soils have not acted upon the surface of
     the piers.

2.   The lower bedrock consists of a combination of sandstone and claystone
     which does not possess a high swell potential. Therefore, the piers
     are relatively stable.

It is possible that these buildings will not suffer severe foundation
movement in the future; however, changing water table and other local
conditions may affect the stability of the structure. For instance, in
Buildings 88, 90 and 92, it is anticipated that severe movement will take
place in the near future. This is because of the high swell potential of
the lower soils and the close proximity of the water table.

Area III

This area is located at the southwest portion of the site where ground
surface is high and the water table elevation is deep. At the northeast
portion of the site, the ground water is well below the bottom of the
piers. The condition of these buildings is relatively good. The floor
slabs have not been replaced, nor have there been any remedial measures
taken.

REMEDIAL MEASURES

Remedial measures depend on the extent of the present damage and can be
best described under the three areas mentioned before:

Area I

In this area, the damage is of such extent that drastic measures should
be taken as recommended:

1. All piers should be cut free from the foundation system so that the entire building will not be associated with the lower bedrock. This is necessary because the source of the problem, for the buildings in this area, is caused by the expansion of the lower bedrock. Since the water table conditions cannot be changed, it is necessary to prevent direct contact of the foundation system with the bedrock.

2. Individual pads should be provided beneath the foundation walls to support the building. The pads should be designed for a maximum soil pressure of 2,500 psf and as much dead load pressure as possible. Since the swelling pressure of the upper soils in this area is about 2,000 psf, there should be little danger of foundation movement due to the swelling of the upper soils.

3. Shims should be provided on top of each pad so that the elevation of the building can be adjusted. The shims should be adjusted with an engineering level immediately after completion of the pads and should be readjusted after a period of six months. The above described underpinning operation can be executed either from inside the basement area or from outside. Since it is necessary in most cases to remove the floor slabs, it will probably be more economical to perform the underpinning operation inside the basement.

4. All floor slabs should be separated from the bearing walls with a positive expansion joint. If effective expansion joints cannot be obtained, a gap of approximately one-half inch should be left all around the slab to insure that the slabs will not bind against the bearing walls. When the basement is finished, this gap can be tiled to prevent dirt entering the gap.

5. The use of interior slab-bearing partition walls should be discouraged. If such is necessary, then slip joints should be provided to insure free partition wall movement. The slip joint should apply to all door frames and staircase walls. It should be emphasized that sheetrock on both sides of the partition wall should also be provided with slip joints.

6. In some cases, it may be necessary to provide a subsurface drainage system around the perimeter of the basement area.

7. It will be necessary to carefully check all sewage and water pipes to insure that no leakage has taken place.

The above remedial measures for Area I are expensive and difficult to carry out, but to insure that no further foundation movement will take place and to eliminate some of the existing damage, such remedial measures are necessary and unavoidable.

Area II

Remedial measures for the buildings in Area II depend greatly upon individual building conditions. In general, the following are recommended:

1. Increase the dead-load pressure exerted on the piers by eliminating a number of piers and increasing the span between the piers. This should be carefully designed and planned by a structural engineer. The dead-load pressure exerted on the piers should be not less than 20,000 psf. The end bearing pressure of the piers should not greatly exceed 40,000 psf, with a skin friction of 4,000 psf for the portion of pier in bedrock.

2. Where existing damage is relatively severe, the entire building should be releveled using shims on top of each pier.

3. Air space beneath the grade beams should be carefully checked for effectiveness. All mushrooms above the piers should be removed.

4. Careful inspection of the condition of the floor slabs should be made. If the slabs are binding against the grade beams, then the slabs should be removed and replaced with an effective joint system. Every effort should be made to avoid the transmission of pressure from the slabs to the foundation walls.

5. Interior slab bearing partition walls and drainage systems should be treated as described under Area I.

Area III

No remedial measures are necessary in this area. The buildings are in relatively good condition and unnecessary alterations in this area should be avoided. However, close observation of foundation movement should be maintained. If it is found at a later date that there are signs of foundation movement, both the structural engineer and the soils engineer should be informed so that they may determine if remedial measures are necessary.

Since the investigation covers as many as 251 units, it is difficult to specify the recommended remedial procedure for each building. The soils engineer should be at the site during the execution of the remedial measures recommended by the structural engineer and the soils engineer.

AN ANATOMY OF A LAWSUIT

GENERAL PROCEDURE FOR GEOTECHNICAL INVESTIGATION

The procedures most geotechnical engineers follow for the investigation of a building site are as follows:

For a large tract of land, where there is a choice of building location, a preliminary soil investigation usually is conducted to provide recommendations as to the most favorable location for the proposed building from a geotechnical standpoint.

When the footprint for the building is established, a final geotechnical report is prepared. The report recommends the foundation type, slab construction, drainage precautions and others. The structural engineer uses the report to design the foundation system, and the architect uses the report for site preparation and landscaping.

Sometimes the structural engineer and/or the architect asks the geotechnical engineer to review the final plans and specifications.

During construction, the owner, or the architect who acts as the representative of the owner, asks the geotechnical engineer to examine the drilling of the pier holes, the placement of fill or the installation of the drain system. The scope of the service is specified in the specifications.

ABSTRACT FROM PRELIMINARY SOIL REPORT

The project is the Colorado State Veterans Nursing Home located at Florence, Colorado. A preliminary geotechnical report was prepared in 1973 for the purpose of determining the most desirable location for construction.

On site selection

The northwestern portion of the site is located in a broad valley where the bedrock is generally deep, and the upper soils do not possess swell potential. This area (north of the trail and west of contour line 5300) in our opinion appears to be the most desirable location for construction as shown on Figure VI-1. The figure also shows the final selected location of the building.

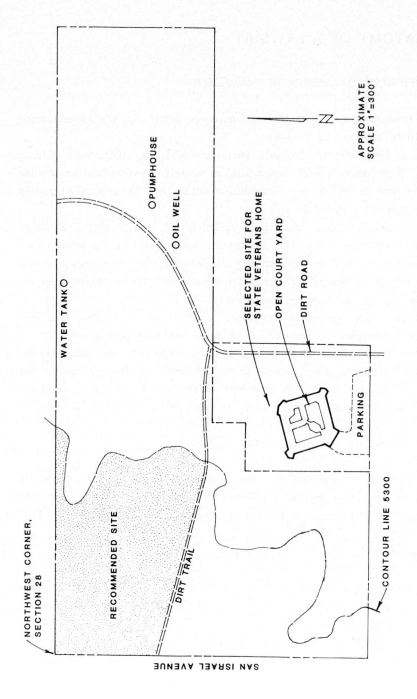

Figure VI-1. Proposed and final location of State Veterans Home.

On foundation system

Where the bedrock is relatively deep and the upper soils do not possess swell potential, spread footing foundations can be used. Where the bedrock is high, such as the area south of the trail, it is desirable to use a drilled pier foundation to support the structure.

On floor slab

Where the upper soils do not possess swelling potential, slab-on-ground construction can be successfully used. However, where bedrock is shallow, interior floor slabs are placed directly on bedrock. There is a strong possibility that heaving soils pose a problem to slab-on-ground construction. Special consideration should be given to slab-on-ground construction to prevent heaving soils from affecting the stability of the structure.

ABSTRACT FROM GEOTECHNICAL REPORT

On foundation system

Considering the subsoil conditions and proposed construction, the most safe and desirable type foundation would be straight-shaft piers drilled into the claystone bedrock. The following design and construction details should be observed:

(1) Piers should be designed for a maximum end pressure of 30,000 psf and a skin friction of 3,000 psf for the portion of pier in the bedrock.

(2) Piers should also be designed for a minimum dead-load pressure of 15,000 psf to resist the potential expansion of the upper soils.

(3) Piers should penetrate a minimum of four feet into bedrock and have a minimum total length of twelve feet.

(4) Piers should be reinforced their full length with at least two #5 bars to resist uplift.

(5) A minimum four-inch void should be provided beneath the grade beam between the piers to prevent uplift on the bottom of the grade beam and to concentrate the load on the piers.

(6) All pier holes should be thoroughly cleaned before placing concrete.

On floor slab

Floor slabs placed on the claystone-sandstone bedrock have a risk of heaving and cracking if the underlying subsoils become wetted. Considering the moderate-to-high expansive potential of the claystone bedrock, it is reasonable to consider a structural floor system with an air space beneath it. Providing the owner is aware of the risk and slab-on-ground floors are required, the following details should be observed:

(1) Floor slabs should be separated from all bearing walls and columns with a positive expansion joint.

(2) Interior partitions resting on the floor slabs should be provided with a slip joint, preferably at the bottom, so that in the event the floor slabs move this movement is not transmitted to the upper structure. This detail is also important for wallboards and door frames.

(3) Floor slabs should be provided with control joints to minimize damage due to shrinkage cracking and they should be reinforced.

(4) A four-inch gravel layer should be placed beneath the floor slabs.

(5) It has been our experience that the risk of floor slab movement can be minimized by removing at least three feet of expansive soil and replacing it with a compacted nonexpansive fill. This should be a granular soil and compacted to at least 90% standard Proctor density at optimum moisture content.

The above precautions do not prevent the movement of floor slabs in the event the floor slabs become wetted; however, they minimize the damage if such movement occurs.

On drainage

The following drainage precautions should be observed during construction and maintained at all times after the building has been completed:

(1) Excessive wetting or drying of the foundation excavation should be avoided during construction.

(2) Backfill around the building should be moistened and compacted to at least 85% standard Proctor density.

(3) The ground surface surrounding the exterior of the building should be sloped to drain away from the building in all directions. We recommend a minimum slope of six inches in the first ten feet.

(4) Roof downspouts and drains should discharge well beyond the limits of all backfill.

(5) Landscaping which requires excessive watering and lawn sprinkler heads should be located at least ten feet from the foundation walls of the building.

Although no ground water was encountered in our test holes at the time of drilling, it is possible that a perched water table condition could occur from time to time due to heavy precipitation or local irrigation. Where the lower floor is below the surface of bedrock, it is possible that this trapped perched water table could seep into the lower floors causing wet conditions. Where the lower floor is below the surface of bedrock, we recommend that an underdrain around the periphery of the building be installed.

DISTRESS AND PUBLICITY

About two years after the completion of the building, it was found that the floor slab had heaved. Upon examination, it was found that a slab-bearing partition wall had not been provided with the necessary slip joints as designed. As the result, not only did the partition wall show cracks, but the walls also pushed against the ceiling and resulted in heavy distortion.

The initial design team advised the administrator to regulate their watering practice, especially in the open courtyard area, and was ignored.

In the meantime, a great deal of publicity was generated, blaming the design engineers and architect for the problem. One paper had the heading, "Chen & Associates charged in negligent construction of Veterans' Home."

SUMMONS

A summons was delivered to the geotechnical engineer in August 1979.

On floor slab

Defendant Chen negligently and improperly recommended slab-on-ground construction with knowledge of the possibility or probability that heaving soil would pose a problem.

The plaintiff, relying on the conclusions and recommendations of Chen for slab-on-ground construction, authorized construction of the Veteran's Nursing Home utilizing slab-on-ground type construction.

As a direct and proximate result of the improper recommendations of Chen to use slab-on-ground construction when the site was not fit for this kind of construction, the plaintiff has been damaged in the manner and in the amounts set forth below.

On pier foundation

Defendant Chen negligently and improperly inspected and supervised the pouring of the concrete piers which resulted in the concrete piers, which are the foundation and support for the Veteran's Nursing Home, not conforming to specifications in that they are larger in diameter than specified.

It was the duty of defendant Chen to notify the contractor of any errors in the pouring of the concrete piers, including errors in the size of said piers. Notwithstanding said duty, defendant Chen improperly failed to notify the contractor of the piers being larger than specified.

On inspection

Defendant Chen improperly failed to notify the contractor that several of the piers had a mushroom or bell formation at the top so that the surface area at the top of the pier was larger than the remainder of the pier and larger than called for by the specifications.

RESPONSE

On floor slabs

Defendant clearly recommended a structural floor system with an air space beneath it rather than slab-on-ground construction. Said report also relates an understanding that the construction will probably be slab-on-ground and states as follows:

"Providing the owner is aware of the risk and slab-on-ground floors are required, we suggest the following details be observed:"

And then there follows five specific recommendations, concluding with the statement that:

"The above precautions will not prevent the movement of floor slabs in the event the floor slabs become wetted; however, they will minimize the damage if such movement occurs."

It is this defendant's contention that the report submitted by the defendant shows on its face that it did not recommend slab-on-ground construction, and, to the contrary, discussed slab-on-ground construction as an alternative with respect to which special precautions and special construction steps had to be observed in order to minimize damage which would have occurred from slab movement if the expansive soil had become wet. It is further this defendant's position that the report made a full disclosure of the dangers of utilizing slab-on-ground construction and this defendant's statements and recommendations contained in the said reports, as a matter of law, do not amount to negligence.

The affiant has since learned that after issuance of the Chen and Associates, Inc., report of July 5, 1974 the architect as well as members of the State Department of Administration and other executive persons of the state of Colorado interested in the project, applied to the appropriate committees of the legislature for additional funding authorization in order to permit the construction of the facility with a structural floor system as recommended in the report of July 5, 1974, but were not successful. Neither the affiant nor the defendant Chen and Associates, Inc., were ever consulted or contacted with respect to those efforts.

On pier size

With respect to the inspection and the supervision function which defendant, Chen & Associates, did undertake, the drilling of the holes for the piers included ascertaining that where an 8" pier was called for in the specifications the hole was 8" in diameter and that all holes were to the proper depth. A letter attached from the structural engineer to the architect stated,

"I received a call from Chen & Associates advising that the contractor was drilling 10" round piers in lieu of 8" round shown on the drawings. As we discussed, it was my understanding that the contractor had asked for a substitution to use 10" round piers. Upon checking, we found that the 10" pier could be substituted for the piers, providing 1'-0" additional minimum penetration into bedrock is provided."

The above letter clearly shows that there was a change in the specifications and the contractor was permitted to substitute 10" diameter piers for the 8" diameter piers.

The structural engineer on the project was charged with the duty of approving or disapproving of such changes. That letter together with the pier inspection reports of the defendant, Chen & Associates, show clearly that all pier holes drilled and approved were approved in accordance with the project specifications as modified and approved by the structural engineer. In addition, it is clearly shown by Mr. Chen's testimony that the increase in the size of the piers which was approved by the structural engineer was still within the minimum dead-load pressure parameter of 15,000 pounds per square foot recommended by the defendant, Chen & Associates, Inc.

On mushroom

The plaintiff alleges that during construction the defendant improperly failed to notify the contractor that several of the piers had a mushroom bell formation at the top so that the surface area at the top of the pier was larger than the remainder of the pier and larger than called for by specifications.

Excerpts from the construction documents which set forth the responsibilities of the soils engineer on the project are as follows:

-   The owner shall employ a soils engineer to inspect the bearing material. If the bearing stratum at design depth is not suitable, the excavation shall be drilled deeper as directed by the soils engineer or the architect.

-   Holes shall be inspected by the architect immediately prior to placing of concrete. Notify architect at least forty-eight hours before beginning the drilling operation.

The above documents clearly show that defendant, Chen & Associates, Inc., undertook to perform only those functions required by the construction documents to be performed by the soils engineer. The only functions relating to the piers performed by the soils engineer pursuant to the construction documents were the inspection and supervision of the drilling of the holes for the piers, which did not include the subsequent separate operation of pouring the piers themselves or placing steel reinforcement in the holes prior to the pouring of the concrete. Thus, defendant, Chen & Associates, Inc., had no duties whatsoever with respect to the inspection of poured piers and consequently no duty to ascertain whether or not mushroom or bell shaped formations appeared thereon after they were poured.

EXPERT TESTIMONY

Throughout the lawsuit, no less than 30 witnesses were called. Among them were five structural engineers, four geotechnical engineers and two architects. Some of the abstract of the testimony is given as follows:

Architect testimony (plaintiff)

The plaintiff engaged an architect from out of state who had no experience on actual construction and nothing on expansive soils. He made the following comments on the geotechnical involvement,

For the preliminary report:
a. In our opinion, the Engineer's report is confusing, erratic and contradictory. It is difficult to understand how a site can be described in the first instance as having no unusual subsoil conditions and in the second ˈinstance described as having erratic bedrock with high swell potential and upper soils that will settle excessively upon wetting.
b. We find the slab-on-ground recommendation very risky on the site. The recommendation is made for situations where soils do not swell. However, according to the report, lower portions of bedrock possess swell potential and upper soils consolidation potential. There are cautions against slab-on-ground construction making the recommendation even more confusing.
c. The recommendation for spread footings, even if on natural soils, and in consideration of the possibility of consolidating soils, in our opinion, is improper."

For the final report:
a. The engineer's statement of Proposed Construction indicates the Architect, presumably based on the first soils report, has designed the building with slab-on-ground.
b. This report generally repeats the caution of slab-on-ground construction, and includes a recommendation for structural floor slabs

with air space under. The report goes on, however, to approve slab-on-ground construction suggesting a risk and recommending a design to minimize damage.

c.   The above situations certainly are confusing and contradictory enough; however, to add to the problem the recommendations on roof drains are short sighted and dangerous if the previous caution given by the Engineer to maintain excavations in a state of equilibrium, i.e., not allowing wetting or drying of foundations excavations, is valid. Roof drainage, if the situation could be critical, should be piped off the site without the possibility of wetting soils.

d.   With the possibility of erratic soil conditions, including moderate-to-high swell potential for claystone bedrock, drilled pier foundations appear to be a risky recommendation. In our opinion, the drilled pier is vulnerable from skin friction and lifting once swelling takes place due to wetting. In addition, if soils absorb moisture, said moisture has access to highest swell area due to drilled piers.

The allegation of the state against the geotechnical engineers appeared to be based entirely on the above testimony. As stated by a structural engineer,

"In general, this report strikes me as a 'quick and dirty' report by a person with little or no experience or understanding of expansive soil foundation problems. This needs to be brought out. The only purpose I can see for the report was to get a lawsuit started."

He states that the soils reports are "confusing, erratic and contradictory." He probably thinks it is erratic and contradictory because the report was for two entirely different building sites which he apparently did not take the time to discover.

He further states that drilled pier foundations appear to be a risky recommendation. This is probably the best statement in the whole report to discredit his opinion. Drilled pier foundations are used very extensively throughout Colorado and, in most cases, they are used most specifically to minimize the effect of expansive soil. For all intents and purposes, there is no other practical foundation system that will accomplish as much to prevent problems with expansive soil. He obviously does not understand the mechanisms of expansive soil and what proper engineering design does to minimize these effects.

Geologist testimony (plaintiff)

The state further sought comment from the state geologist and their comments on the geotechnical engineer are as follows:

1. The owner accepted a degree of risk when the site with its expansive soils was selected, in light of known problems with other state buildings in other areas on expansive soils.

2. The soils engineer identified the expansive characteristics of the soils, recommended ways to cope with these soils and how to keep them dry but I feel did not emphasize strongly enough the dire consequence of poor roof and site drainage.

3. The design of the structure with its open courtyards, lack of positive interior roof drainage, lack of positive outer lawn drainage is highly at fault as to the long-term performance of the foundation and slab system. The fact that these problems were not identified at the time of construction indicates that there was no responsible individual on the job who knew the consequence of these design and construction deficiencies.

4. A structural floor slab system was the first recommendation with no stated risk. Slab-on-ground floors could be used if certain details were observed but being clearly stated that such floors could heave and crack if the underlying subsoils became wet. In other words, even if the stated precautions were observed, if the soils under the slab became wet, the floor slab would move.

The plaintiff saw that the state geologist did not find any fault on the geotechnical engineer, made a strong point that, "The geotechnical engineer did not emphasize strongly enough the dire consequence of poor roof and site drainage." This later became the central target of the state against the defendant.

Structural engineer testimony (defendant)

A structural engineer who has vast experience on structural damage on expansive soil reported as follows:

"It is also my opinion the recommendations in the Chen report are relatively standard for situations of this nature and are in accord with normal engineering practice.

"Some of the prior investigative reports indicated the possibility of grade beam and foundation movement. However, at the time of my inspection there was little obvious evidence of foundation movement, although no actual measurements were made.   The building damage consisted primarily of cracks and other distress that had occurred in the drywall partitions, doorways, ceilings and floors.   It appeared most and possibly all of this damage was attributable to movement of the concrete floor slab which in turn was causing movement of the partitions built on it.

"Considering the highly expansive nature of the supporting soil, the extremely poor surface drainage on all sides of the building, the excessive irrigation that has been reported, and the potential for really severe problems, the building was in relatively good condition.   Most of the cracking should be considered minor and no serious structural problems were noted.

"Slab-on-ground   construction   is   not   a   'negligent   or   improper' recommendation.   Buildings are designed and constructed on expansive soils using slab-on-ground construction constantly throughout Colorado.   Very few building owners are willing to spend the substantial extra money necessary to eliminate slab-on-ground construction.

"The increase in drilled pier (or caisson) diameter was a properly approved design change that occurred during construction."

This report was not considered seriously by the state.   They claim that local engineers tend to protect fellow engineers.

Academician testimony (plaintiff)

A college professor from Chicago provided a geotechnical report.   He first performed an analysis on the clay mineral content of the claystone shale and concluded:
1.   The clay-shale foundation rock has a potential to and did swell.
2.   The flare and the enlarged pier diameter allow more force from the heaving shale to be transferred to the pier than would original design.
3.   The crack at the bottom shows that the pier did not provide sufficient anchorage to counteract the forces created by the swelling shale."

It is interesting to note that the crack developed on the pier can only indicate that there is sufficient anchorage of the pier.

Structural engineer testimony (defendant)

In direct response to the academician's report, the structural engineer stated:

1. The change from eight-inch diameter piers to ten-inch diameter piers was also checked, and we found that this met the design criteria as approved by the original design engineer.
2. The mushrooming (flare) at the top of the caissons is not termed as critical. It is the opinion of an independent geotechnical engineer that the additional anchorage into the claystone bedrock more than offsets the uplift on the additional small area created by the mushroom effect.

Geotechnical expert testimony (defendant)

The state of Colorado finally sought testimony from a top geotechnical engineer in the field of expansive soil from Washington, D.C. The state expected that this witness could discredit the geotechnical engineer's report. To their surprise, the expert fully agreed with Chen & Associates' report and actually blamed the state for over-irrigation.

DAMAGE TO THE PROFESSION

The case started in August, 1979, and after lengthy deposition and discovery was finally settled in December, 1983 after 40 months of frustration.

The claim and the settlement

The state sued the architect, the general contractor and the geotechnical engineer for a total of 1.5 million dollars. Considering the cost of the entire building was only slightly over 1.0 million dollars, it has to be as the state witness stated, "a total loss." In fact, the building has not suffered a single day of loss to its occupancy. The 1.5 million dollar claim includes the provision of a new TV set for each resident unit and a new fence around the site.

The cost of this lawsuit involved not only the settlement cost but also the cost of attorney fees, time spent in deposition, preparation and others. A rough estimate on the total cost from the defendant is as follows:

	Out of Pocket Cost	Settlement Cost
Contractor	30,000	40,000
Architect	45,000	30,000
Geotechnical engineer	47,000	0
Total	192,000	

Of course, the cost to the plaintiff on this long affair was borne by the taxpayer.

Of the total, 47,000 spent by the geotechnical engineer in the span of 40 months, the firm has received $2,057 for a geotechnical report and $3,425 for field observation.

The truth of the problem

The truth of the problem has never been fully revealed throughout the long deposition. The facts are as follows:

1. The movement was mainly from the heaving of the slab-on-ground.
2. Slab bearing partition walls pushed the ceiling and caused most of the distress.
3. The source of moisture was from the watering of the courtyard area. No adequate drainage has been provided in the courtyards.
4. The piers have not been lifted except for the patio area where the piers are lightly loaded.
5. After the watering of the courtyard area has stopped, the floor actually settled back. The three feet of structural fill beneath the slab actually helped to minimize damage from otherwise more severe differential heaving.

Conditional suspension

In the meantime, the state issued the following letter to all concerned, blacklisting the geotechnical engineer from participating in any further state projects:

"Because of Chen & Associates' performance in the construction of the Colorado State Veterans Nursing Home at Florence, Colorado, serious questions have arisen as to the ability of Chen & Associates to perform in a first class, substantial manner on a major construction project. This matter is now in litigation.

"Therefore, the Department has qualified Chen & Associates on the following basis. Chen & Associates are qualified to perform services on state construction projects, although this qualification will be reviewed in light of work Chen & Associates has performed, is now performing, or may in the future, perform for the state. Pursuant to statute, should Chen & Associates be considered for providing consulting engineering services on any construction project, the Department will determine whether it is appropriate in that project situation to recommend Chen & Associates for the work.

"Chen & Associates should be advised that its work for the state will be scrutinized to determine whether Chen & Associates is performing its contract in accord with the intent and requirements of the contract."

The building was repaired at a low cost and is now performing well. Cost of remedial construction could be much lower, if the state chose to follow the recommendations of the geotechnical engineer instead of initiating the lengthy lawsuit.

Has justice been served in this entire scenario? Should geotechnical engineers risk their reputation in dealing with expansive soil?

APPENDIX

APPENDIX

APPENDIX A

Standard Test Methods for

ONE-DIMENSIONAL SWELL OR SETTLEMENT POTENTIAL OF COHESIVE SOILS[1]

This standard is issued under the fixed designation D 4546; the number immediately following the designation indicates the year of original adoption or, in the case of revision, the year of last revision. A number in parentheses indicates the year of last reapproval. A superscript epsilon (ϵ) indicates an editorial change since the last revision or reapproval.

1. Scope

1.1 These test methods cover three alternative laboratory methods for determining the magnitude of swell or settlement of relatively undisturbed or compacted cohesive soil.

NOTE 1—Refer to Section 5 to determine the best method for a particular application.

1.2 The methods can be used to determine (a) the magnitude of swell or settlement under known vertical (axial) pressure, or (b) the magnitude of vertical pressure needed to maintain no volume change of laterally constrained, axially loaded specimens.

1.3 The values stated in SI units are to be regarded as the standard. The values stated in inch-pound units are approximate.

1.4 *This standard may involve hazardous materials, operations, and equipment. This standard does not purport to address all of the safety problems associated with its use. It is the responsibility of whoever uses this standard to consult and establish appropriate safety and health practices and determine the applicability of regulatory limitations prior to use.*

2. Applicable Documents

2.1 *ASTM Standards:*

D 422 Method for Particle-Size Analysis of Soils[2]

D 653 Terms and Symbols Relating to Soil and Rock[2]

D 698 Test Methods for Moisture-Density Relations of Soils and Soil-Aggregate Mixtures Using 5.5-lb (2.49-kg) Rammer and 12-in. (305-mm) Drop[2]

D 854 Test Method for Specific Gravity of Soils[2]

D 1557 Test Methods for Moisture-Density Relations of Soils and Soil-Aggregate Mixtures Using 10-lb (4.54-kg) Rammer and 18-in. (457-mm) Drop[2]

D 1587 Practice for Thin-Walled Tube Sampling of Soils[2]

D 2216 Method for Laboratory Determination of Water (Moisture) Content of Soil, Rock, and Soil-Aggregate Mixtures[2]

D 2435 Test Method for One-Dimensional Consolidation Properties of Soils[2]

D 3550 Practice for Ring-Lined Barrel Sampling of Soils[2]

D 3877 Test Methods for One-Dimensional Expansion, Shrinkage, and Uplift Pressure of Soil-Lime Mixtures[2]

D 4220 Practices for Preserving and Transporting Soil Samples[2]

D 4318 Test Method for Liquid Limit, Plastic Limit, and Plasticity Index of Soils[2]

3. Terminology

3.1 *Definitions*—Refer to Terms and Symbols D 653 for standard definitions of terms.

3.2 *Descriptions of Terms Specific to This Standard:*

3.2.1 *heave (L)*—increase in vertical height, Δh, of a column of in situ soil of height h following sorption of water.

3.2.2 *percent heave or settlement, %*—increase or decrease in the ratio of the change in vertical height, Δh, to the original height of a column of

---

[1] These test methods are under the jurisdiction of ASTM Committee D-18 on Soil and Rock and are the direct responsibility of Subcommittee D18.05 on Structural Properties of Soils.

Current edition approved Nov. 29, 1985. Published January 1986.

[2] *Annual Book of ASTM Standards*, Vol. 04.08.

in situ soil; $h \times 100$ or $\Delta h/h \times 100$.

3.2.3 *settlement, L*—decrease in vertical height, Δh, of a column of in situ soil of height h.

3.2.4 *swell, L*—increase in elevation or dilation of soil column following sorption of water.

3.2.5 *free swell, %*—percent heave, $\Delta h/h \times 100$, following sorption of water at the seating pressure σ_{se}.

3.2.6 *primary swell, L*—an arbitrary short-term swell usually characterized as being completed at the intersection of the tangent of reverse curvature to the curve of a dimensional change-logarithm of time plot with the tangent to the straight line portion representing long-term or secondary swell (Fig. 1).

3.2.7 *secondary swell, L*—an arbitrary long-term swell usually characterized as the linear portion of a dimensional change-logarithm of time plot following completion of short-term or primary swell (Fig. 1).

3.2.8 *swell index*—slope of the rebound pressure - void ratio curve on a semi-log plot.

3.2.9 *swell pressure, FL^{-2}—(1)* a pressure which prevents the specimen from swelling as obtained in Method C, or *(2)* that pressure which is required to return the specimen back to its original state (void ratio, height) after swelling in Method A or B.

Note 2—Swell pressures by Method C corrected for specimen disturbance may be similar to or slightly greater than those by Method A.

4. Summary of Methods

4.1 The following three alternative methods require that a soil specimen be restrained laterally and loaded axially in a consolidometer with access to free water.

4.1.1 *Method A*—The specimen is inundated and allowed to swell vertically at the seating pressure (pressure of at least 1 kPa (20 lbf/ft²) applied by the weight of the top porous stone and load plate) until primary swell is complete. The specimen is loaded after primary swell has occurred until its initial void ratio/height is obtained.

4.1.2 *Method B*—A vertical pressure exceeding the seating pressure is applied to the specimen before placement of free water into the consolidometer. The magnitude of vertical pressure is usually equivalent to the in situ vertical overbur-

den pressure or structural loading, or both, but may vary depending on application of the test results. The specimen is given access to free water. This may result in swell, swell then contraction, contraction, or contraction then swell. The amount of swell or settlement is measured at the applied pressure after movement is negligible.

4.1.3 *Method C*—The specimen is maintained at constant height by adjustments in vertical pressure after the specimen is inundated in free water to obtain swell pressure. A consolidation test is subsequently performed in accordance with Test Method D 2435. Rebound data is used to estimate potential heave.

5. Significance and Use

5.1 The relative swell/settlement potential of soil determined from these test methods can be used to develop estimates of heave or settlement for given final moisture and loading conditions. The initial water content and void ratio should be representative of the in situ soil immmediately prior to construction. Selection of test method, loading, and inundation sequences should, as closely as possible, simulate any construction and post-construction wetting and drying effects and changes in loading conditions.

5.2 Soils containing montmorillonites (Smectite) are likely to have a significant potential for swell and are commonly tested by these test methods.

Note 3—Montmorillonites with divalent cations usually swell less than with monovalent cations. It is useful to know the type of cation as well as the cation exchange capacity of montmorillonite.

5.3 Laboratory-prepared test specimens should duplicate the in situ soil or field-compacted soil conditions as closely as possible because relatively small variations in unit weight and water content can significantly alter the measured heave and swell pressure. Differences in soil fabric of the compacted specimens, such as obtained by kneading or static compaction, could also have a significant impact on the swell/settlement behavior of cohesive soils.

5.4 These test methods are applicable to undisturbed test or remolded specimens, or both, as follows:

5.4.1 *Method A*—This test method measures *(a)* the free swell, *(b)* percent heave for vertical confining pressures up to the swell pressure, and

(c) the swell pressure.

5.4.2 *Method B*—This test method measures (a) the percent heave or settlement for vertical pressure usually equivalent to the estimated in situ vertical overburden and other vertical pressure up to the swell pressure, and (b) the swell pressure.

5.4.3 *Method C*—This test method measures (a) the swell pressure, (b) preconsolidation pressure, and (c) percent heave or settlement within the range of applied vertical pressures.

NOTE 4—Methods A and C have produced estimates of heave consistent with observed heave. Method B may lead to estimates of heave less than observed heave. Method A has not been recommended for evaluation of swell pressure and consolidation parameters for settlement estimates because sorption of water under practically no restraint may disturb the soil structure.

6. Interferences

6.1 Estimates of the swell and settlement of soil determined by these test methods are often of key importance in design of floor slabs on grade and evaluation of their performance. However, when using these estimates it is recognized that swell parameters determined from these test methods for the purpose of estimating in situ heave of foundations and compacted soils may not be representative of many field conditions because:

6.1.1 Lateral swell and lateral confining pressure are not simulated.

6.1.2 Swell in the field usually occurs under constant overburden pressure, depending on the availability of water. Swell in the laboratory is evaluated by observing changes in volume due to changes in applied pressure while the specimen is inundated with water. Method B is designed to avoid this limitation.

6.1.3 Rates of swell indicated by swell tests are not always reliable indicators of field rates of heave due to fissures in the in situ soil mass and inadequate simulation of the actual availability of water to the soil. The actual availability of water to the foundation may be cyclic, intermittent, or depend on in-place situations, such as pervious soil-filled trenches and broken water and drain lines.

6.1.4 Secondary or long-term swell may be significant for some soils and should be added to primary swell.

6.1.5 Chemical content of the inundating water affects volume changes and swell pressure; that is, field water containing large concentrations of calcium ions will produce less swelling than field water containing large concentrations of sodium ions or even rain water.

6.1.6 Disturbance of naturally occurring soil samples greatly diminishes the meaningfulness of the results.

7. Apparatus and Materials

7.1 *Consolidometer*—The apparatus shall comply with the requirements of Test Method D 2435. The apparatus shall be capable of exerting a pressure on the specimen of (1) at least 200 % of the maximum anticipated design pressure, or (2) the pressure required to maintain the original specimen height when the specimen is inundated (Method C), whichever is greatest.

7.1.1 Consolidometer rigidity influences the observed swell, particularly with Method C. Therefore, consolidometers of high rigidity should be used with Method C (see Test Method D 2435).

NOTE 5—Small increases in soil volume can significantly relieve swell pressures. Therefore, variations in displacements that occur during determination of swell pressures by Method C should be as small as possible to reduce the magnitude of correction required in 13.2.5. The measurements, especially swell pressure measurements, should be based on corrections for compression of members.

7.2 *Porous Stones*—The stones shall be smooth ground and fine enough to minimize intrusion of soil into the stones if filter paper is not used and shall reduce false displacements caused by seating of the specimen against the surface of porous stones (Note 6). Such displacements may be significant, especially if displacements and applied vertical pressures are small.

7.2.1 Porous stones shall be air dry.

7.2.2 Porous stones shall fit close to the consolidometer ring to avoid extrusion or punching at high vertical pressures. Suitable stone dimensions are described in 5.3 of Test Method D 2435.

NOTE 6—A suitable pore size is 10 μm if filter paper is not used. Filter paper is not recommended because of its high compressibility and should not be used when measuring the swell/settlement of stiff clays and when measuring swell pressure by Method C.

7.3 *Plastic Membrane, Aluminum Foil, or Moist Paper Towel*, a loose fitting cover to enclose the specimen, ring, and porous stones prior to inundating the specimen, used to minimize evaporation from the specimen.

8. Sampling of Naturally Occurring Soils

8.1 Disturbance of the soil sample from which specimens are to be obtained greatly diminishes the meaningfulness of results and should be minimized. Practice D 1587 and Practice D 3550 cover procedures and apparatus that may be used to obtain satisfactory undisturbed samples.

8.2 Storage in sampling tubes is not recommended for swelling soils even though stress relief may be minimal. The influence of rust and penetration of drilling fluid or free water into the sample may adversely influence laboratory test results. Water and oxygen from the sample could cause the formation of rust within the tube which could result in the sample adhering to the tube. Therefore, sampling tubes should be brass, stainless steel, or galvanized or lacquered inside to inhibit corrosion in accordance with Practice D 1587.

8.3 If samples are to be stored prior to testing, they should be extruded from the sampling tubes as quickly as possible after sampling and thoroughly sealed to minimize further stress relief and moisture loss. The sample should be extruded from the sampling tube in the same direction as sampled, to minimize further sample disturbance. If the sample cannot be extruded from the tubes immediately, they should be handled and shipped in accordance with Practices D 4220, Group D.

8.4 Prior to sealing in storage containers, samples extruded from tubes that were obtained with slurry drilling techniques should be wiped clean to remove drilling fluid adhering to the surface of the sample. An outer layer of 3 to 6 mm (0.1 to 0.3 in.) should be trimmed from the cylindrical surface of the samples so that moisture or the slurry will not penetrate into the sample and alter the swell potential, swell pressure, and other soil parameters. Such trimming will also remove some disturbance at the periphery due to sidewall friction. Drilling with air or foam instead of slurry will reduce moisture penetration.

8.5 Containers for storage of extruded samples may be either cardboard or metal and should be approximately 25 mm (1 in.) greater in diameter and 40 to 50 mm (1.5 to 2.0 in.) greater in length than the sample to be encased.

8.6 Soil samples stored in containers should be completely sealed in wax. The temperature of the wax should be 8 to 14°C (15 to 25°F) above the melting point when applied to the soil sample; wax that is too hot will penetrate pores and cracks in the sample and render it useless and will also dry the sample. Aluminum foil, cheese cloth, or plastic wrap may be placed around the sample to prevent penetration of molten wax into open fissures. A small amount of wax (about 113-mm or 0.5-in. thickness) should be placed in the bottom of the container and allowed to partly congeal. The sample should subsequently be placed in the container, completely immersed and covered with molten wax, and then allowed to cool before moving.

NOTE 7—A good wax for sealing expansive soils consists of a 1 to 1 mixture of paraffin and microcrystalline wax or 100 % beeswax.

8.7 Examine and test samples as soon as possible after receipt; however, samples required to be stored should be kept in a humid room and may require rewaxing and relabeling before storage. Samples encased in wax or sampling tubes may be cut using a band-saw. The soil specimen should be adequately supported while trimming to size using sharp and clean instruments. The specimen may be extruded from a section of sampling tube and trimmed in one continuous operation to minimize sampling disturbance.

9. Specimen Preparation

9.1 Undisturbed or laboratory-compacted specimens may be used for testing. Prepare laboratory-compacted specimens to duplicate compacted fills as closely as possible.

NOTE 8—The compaction method, such as kneading or static compaction, may influence the volume change behavior when prepared wet of optimum water content. Compaction of laboratory specimens is described in Test Methods D 698 and Test Methods D 1557. Swelling soil is sometimes adequately treated with lime and test specimens compacted as described in Test Methods D 3877.

9.2 Trim the specimen in accordance with Test Method D 2435. A ring extension or guide ring as shown in Test Methods D 3877 may be added to the consolidometer assembly to accommodate specimen swell. Alternatively, a thin hard disk may be inserted in the bottom of the specimen ring during compaction or trimming of a specimen into the ring. Turn the ring and specimen upside down and remove the thin disk insert

to provide space for specimen swell. Take precaution to minimize disturbance of the soil or
changes in moisture and unit weight during sample transportation and preparation. Vibration,
distortion, and compression must be avoided.

NOTE 9—Tests with specimens recessed 5 mm (0.2
in.) in rings of 25-mm (1.0-in.) height have performed
adequately.

10. Calibration

10.1 Calibrate the consolidation machine in
accordance with Test Method D 2435.

10.2 Measure the compressibility of the apparatus with a smooth copper, brass, or hard steel
disk substituted for the soil specimen. The disk
should be approximately the same height as the
specimen and 1 mm (0.04 in.) smaller in diameter than that of the ring. Place moistened filter
papers between the porous stones and metal disk
if filter papers are to be used during the test.
Allow sufficient time for moisture to be squeezed
from the filter paper during each load increment
and decrement. The deflections of the calibration
test are subtracted from the deflections of the soil
test for each load increment and decrement.

NOTE 10—When filter paper is used, calibration
must duplicate the exact load increment/decrement
sequence due to inelastic compression of paper; thus,
calibration is needed for each test. Periodic calibration
will suffice for tests without filter paper.

11. Associated Soil Properties

11.1 Determine the initial (or natural) water
content, wet and dry unit weights, volume, and
initial void ratio in accordance with Test Method
D 2435. Determine the specific gravity in accordance with Test Method D 854 when results are
required in terms of void ratio. The liquid limit,
plastic limit, and plasticity index as determined
in accordance with Test Method D 4318 and the
particle size distribution for soils with substantial
granular material as determined in accordance
with Method D 422 are useful in identifying the
soil and correlating results of tests on different
soils.

12. Procedure

12.1 Assemble the ring with the specimen recessed in the ring, dry filter paper if used, and
air-dry porous stones in the loading device. Enclose the specimen, ring, filter paper, if any, and
porous stones as soon as possible with a loose
fitting plastic membrane, moist paper towel, or

aluminum foil to minimize change in specimen
water content and volume due to evaporation.
This wrapping may be cut away and discarded at
the time of specimen inundation.

12.2 Apply a seating pressure, σ_{se}, of at least 1
kPa (20 lbf/ft^2). Within 5 min after application
of σ_{se}, adjust the extensometer deformation device for the initial or zero reading.

12.3 A graphical representation of results of
the three alternative test methods shown in Fig.
2 includes corrections for consolidometer compressibility. These test methods are performed in
accordance with Test Method D 2435 except as
follows:

12.3.1 *Method A*—After the initial deformation reading at the seating pressure is recorded,
inundate the specimen and record deformations
after various elapsed times. Readings at 0.1, 0.2,
0.5, 1.0, 2.0, 4.0, 8.0, 15.0, and 30.0 min and 1,
2, 4, 8, 24, 48, and 72 h are usually satisfactory.
Continue readings until primary swell is complete, as determined by the method illustrated in
Fig. 1. After completion of swell, apply a vertical
pressure of approximately 5, 10, 20, 40, 80, etc.,
kPa (100, 200, 400, 800, 1600, etc., lbf/ft^2) with
each pressure maintained constant in accordance
with 10.4 of Test Method D 2435. Maintain pressure until the specimen is recompressed to its
initial void ratio/height. The duration of each
load increment shall be equal and of a duration
which assures 100 % primary consolidation (see
11.2 or 11.6 of Test Method D 2435).

NOTE 11—Some secondary swell must be recorded
in order to determine graphically the end of primary
swell.
NOTE 12—The duration of a typical loading increment is 1 day.
NOTE 13—Vertical pressures may be applied to re-
compress the specimen to void ratios less than the initial
void ratio (point 6, Fig. 2 (Method A)) because the
exact magnitude of vertical pressure required to recom-
press the specimen to its initial void ratio is unknown.
Loading units equipped with pneumatic regulators are
ideally suited for this purpose.

12.3.2 Method A may be modified to place
an initial vertical stress, σ_1, on the specimen
equivalent to the estimated vertical pressure on
the in situ soil within 5 min of placing the seating
pressure and securing the zero deformation reading. Read the deformation within 5 min and
remove the vertical stress, except for the seating
pressure. Record the deformation within 5 min
after removal of σ_1, inundate the specimen, and
continue the test as in 12.3.1. This modification

provides a correction to the initial deformation reading at σ_{se} in an effort to more closely duplicate the in situ void ratio of the soil.

12.3.3 *Method B*—Apply a vertical pressure exceeding the seating pressure within 5 min of placing the seating pressure. Read the deformation within 5 min of placing the vertical pressure. The specimen is inundated immediately after the deformation is read and deformation recorded after elapsed times similar to 12.3.1 until primary swell is complete. Continue the test as in 12.3.1.

12.3.4 *Method C*—Apply an initial stress, σ_1, equivalent to the estimated vertical in situ pressure or swell pressure within 5 min after placement of the seating pressure. Read the deformation within 5 min after placing σ_1, and immediately inundate the specimen with water. Apply increments of vertical stress as needed to prevent swell (see Note 14). Variations from the deformation reading at the time the specimen is inundated at stress σ_1 shall be kept preferably within 0.005 mm (0.0002 in.) and not more than 0.010 mm (0.0004 in.). Load the specimen in accordance with 12.3.1, following no further tendency to swell (usually overnight). Load increments shall be sufficient to define the maximum point of curvature on the consolidation curve and to determine the slope of the virgin compression curve. The rebound curve following consolidation shall also be determined as illustrated in Fig. 2 (Method C). Duration of rebound load decrements shall be in accordance with 10.6 of Test Method D 2435.

NOTE 14—The use of small weight increments, such as lead shot, provide adequate control as needed to prevent swell.

12.4 Measurements shall include the time of reading, applied stress, observed deformation, and corrections for compression of members.

13. Calculations

13.1 Compute the initial void ratio or height, water content, wet and dry unit weights, and degree of saturation in accordance with Test Method D 2435. The void ratio or percent heave calculations are based on the final dial reading for each swell increment and load increment or decrement. The void ratio or percent heave may be plotted versus logarithm of the vertical pressure, as for examples of the three methods graphically illustrated in Fig. 2. The percent heave shall be relative to an initial specimen height, h_o,

observed for an appropriate applied vertical pressure, σ (see 4.1.2). Void ratio or percent heave versus vertical pressure on an arithmetic scale may also be useful for practical applications.

13.2 The data points from a plot of e versus $\log_{10} \sigma$ (Fig. 2) may be used to evaluate the swell and settlement parameters of the tested soil.

13.2.1 *Method A*—The free swell at the seating pressure relative to the initial void ratio, e_o, is given as follows (see Fig. 2 (Method A)):

$$\frac{\Delta h}{h_o} \times 100 = \frac{e_{se} - e_o}{1 + e_o} \times 100 = \left(\frac{\gamma_{do}}{\gamma_{dse}} - 1\right) 100$$

where:
Δh = change in specimen height,
h_o = initial specimen height,
e_{se} = void ratio after stabilized swell at the seating pressure σ_{se},
e_o = initial void ratio,
γ_{do} = dry unit weight at void ratio e_o, and
γ_{dse} = dry unit weight at void ratio e_{se}.

NOTE 15—Figure 2 (Method A) illustrates the free swell at a seating pressure $\sigma_{se} = 1$ kPa (20 lbf/ft^2).

$$\frac{\Delta h}{h_o} \times 100 = \frac{0.908 - 0.785}{1.000 + 0.785} \times 100 = 6.9\ \%$$

The percent heave of 6.9 % may be read directly from the right ordinate of Fig. 2 (Method A) for $e_{se} = 0.908$, point 4.

13.2.2 The percent heave at a vertical pressure, σ, up to the swell pressure σ_{sp}, relative to e_o or an appropriate initial vertical pressure σ_{vo}, is as follows (see Fig. 2 (Method A)):

$$\frac{\Delta h}{h_o} \times 100 = \frac{e - e_o}{1 + e_o} \times 100 = \left(\frac{\gamma_{do}}{\gamma_d} - 1\right) 100$$

where:
e = void ratio at vertical pressure, and
γ_d = dry unit weight at void ratio e.

NOTE 16—Figure 2 (Method A) illustrates a percent heave, as follows:

$$\frac{\Delta h}{h_o} \times 100 = \frac{0.830 - 0.785}{1.000 + 0.785} \times 100 = 2.5\ \%$$

where:
$e = e_{vo} = 0.830$, and
$\sigma = \sigma_{vo} = 100$ kPa (2000 lbf/ft^2).
The swell pressure, σ_{sp}, is given by 400 kPa (8350 lbf/ft^2) relative to $e_o = 0.785$.

13.2.3 Figure 2 may be plotted with dry unit weight, γ_d, versus logarithm of applied pressure, σ, instead of void ratio e versus logarithm σ if specific gravities were not determined. The swell for any change in dry unit weight within limits of the test results may be determined in a manner similar to that described in 13.2.1.

13.2.4 *Method B*—The percent heave at the vertical pressure σ_{vo}, applied following the seating pressure, (see 4.1.2) relative to e_o is given as follows (see Fig. 2 (Method B)):

$$\frac{\Delta h}{h_o} \times 100 = \frac{e_{vo} - e_o}{1 + e_o} \times 100 = \left(\frac{\gamma_{do}}{\gamma_{dvo}} - 1\right) 100$$

where:

e_{vo} = void ratio after stabilized swell at the applied vertical pressure σ_{vo}, and

γ_{dvo} = unit dry weight at void ratio e_{vo}.

NOTE 17—Figure 2 (Method B) illustrates a percent heave, as follows:

$$\frac{\Delta h}{h_o} \times 100 = \frac{0.820 - 0.785}{1.000 + 0.785} \times 100 = 2.0\ \%$$

where

$\sigma = \sigma_{vo} = 100$ kPa (2000 lbf/ft²), and
σ_{sp} = swell pressure = 350 kPa (7300 lbf/ft²) for e_o = 0.785.

Computations of settlement are similar if the specimen contracts at the applied vertical pressure following access to water.

13.2.5 *Method C*—The swell pressure σ_{sp} (point 3, Fig. 2 (Method C)) shall be corrected upward by a suitable construction procedure. Soil disturbance and the process of adjusting vertical pressures may allow some volume expansion to occur, which reduces the maximum observed swell pressure.

NOTE 18—Suitable correction procedures include those based on the preconsolidation pressure σ_{vm}. A construction procedure for soils that break onto a "virgin compression" curve when the recompression curve is not apparent is as follows: (*a*) locate the point of maximum curvature (point 5, Fig. 2 (Method C)), (*b*) draw horizontal, tangential, and bisector lines through the point of maximum curvature, (*c*) draw the virgin part of the compression curve backward to intersect the bisector at the preconsolidation pressure σ_{vm}, or 780 kPa (Fig. 2 (Method C)). The swell pressure is taken as the preconsolidation pressure. The slope of the rebound curve of these soils is usually much less than that of the compression curve.

NOTE 19—A modified construction procedure may be used for soils that break onto the recompression curve, Fig. 2 (Method C). The construction procedure is as follows: (*a*) locate the point of maximum curvature (point 4, Fig. 2 (Method C)), (*b*) draw horizontal, tangential, and bisector lines through the point of maximum curvature, (*c*) extend the recompression line through the bisector line. Intersection of the recompression line with the bisector line is designated the corrected swell pressure, σ'_{sp}, which is 380 kPa for the example in Fig. 2 (Method C). A detail of this construction is shown in Fig. 3. σ'_{sp} in this case is less than σ_{vm}. If the recompression line is not well defined, draw a line parallel with the rebound curve for void ratios

greater than e_o through the bisector line. Frequent load increments may be necessary to define any recompression curve.

13.2.6 Draw a suitable curve parallel with the rebound (or recompression) curve for void ratios greater than e_o through the corrected swell pressure σ'_{sp} at the initial void ratio e_o given by point 3, Fig. 2 (Method C), to obtain the percent heave for any vertical pressure relative to σ'_{sp} and e_o within the range of test results.

NOTE 20—Percent heave calculated by Method C for $\sigma_{vo} = 100$ kPa (2000 lbf/ft²) is as follows:

$$\frac{\Delta h}{h_o} \times 100 = \frac{e_{vo} - e_o}{1 + e_o} \times 100$$

$$= \frac{0.828 - 0.785}{1.000 + 0.785} \times 100 = 2.4\ \%$$

13.2.7 The percent settlement (negative percent heave) may be evaluated from the void ratio e_2 exceeding the corrected swell pressure, as follows:

$$\frac{\Delta h}{h_o} \times 100 = \frac{e_2 - e_o}{1 + e_o} \times 100$$

NOTE 21—Figure 2 (Method C) illustrates the percent settlement, as follows:

$$\frac{\Delta h}{h_o} \times 100 = \frac{0.671 - 0.785}{1.000 + 0.785} \times 100 = -6.4\ \%$$

where:

$e_2 = 0.671$, and
$\sigma_2 = 2560$ kPa (53 000 lbf/ft²).

14. Report

14.1 The report shall include the information required in Test Method D 2435, and shall also include the following:

14.1.1 All departures from procedures, including changes in loading sequences.

14.1.2 The percent heave or settlement for the given vertical pressure and swell pressure σ_{sp}, or corrected swell pressure σ'_{sp}. The compression index, C_c, and swell index, C_s, should be reported if these are evaluated. All departures from the described procedures for computing these parameters and correction procedures used to determine percent heave or settlement and σ'_{sp} shall be described.

14.1.3 The type of water used to inundate the specimen.

15. Precision and Bias

15.1 The variability of soil and the resultant

inability to determine a true reference value prevent development of a meaningful statement of
bias. Data are being evaluated to determine the

precision of these test methods. In addition, the
subcommittee is seeking pertinent data from
users of these test methods.

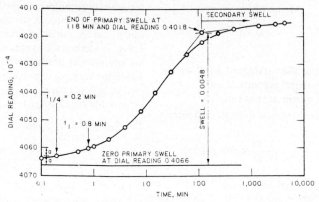

FIG. 1   Time - Swell Curve

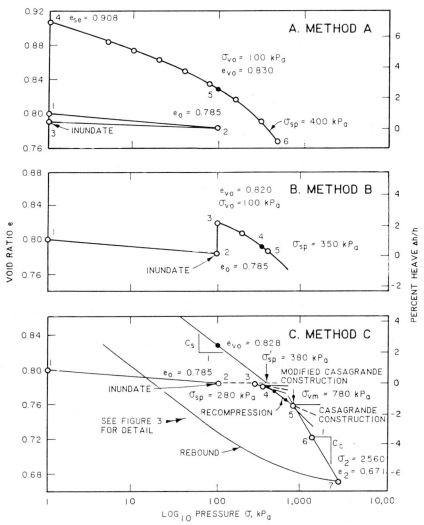

FIG. 2 Void Ratio - Log Pressure Curves

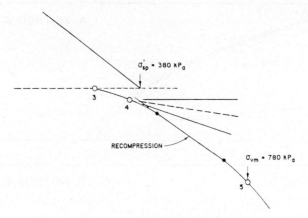

FIG. 3   Construction Detail for Method C

The American Society for Testing and Materials takes no position respecting the validity of any patent rights asserted in connection with any item mentioned in this standard. Users of this standard are expressly advised that determination of the validity of any such patent rights, and the risk of infringement of such rights, are entirely their own responsibility.

This standard is subject to revision at any time by the responsible technical committee and must be reviewed every five years and if not revised, either reapproved or withdrawn. Your comments are invited either for revision of this standard or for additional standards and should be addressed to ASTM Headquarters. Your comments will receive careful consideration at a meeting of the responsible technical committee, which you may attend. If you feel that your comments have not received a fair hearing you should make your views known to the ASTM Committee on Standards, 1916 Race St., Philadelphia, PA 19103.

APPENDIX B

To Convert	Multiply By	To Obtain
cubic feet	2.832×10^{-2}	cubic meters*
cubic feet	7.4805	gallons (U.S. Liquid)
cubic feet	2.832×10^{1}	liters
cubic inches	1.639×10^{1}	cubic centimeters*
farenheit (degrees)-32	5/9	centrigrade (degrees*)
feet	3.048×10^{1}	centimeters*
feet	3.048×10^{-1}	meters*
feet of water	4.335×10^{-1}	pounds/sq. in.
feet of water	3.048×10^{2}	Kgs/sq. meter
gallons	3.785×10^{3}	cubic centimeters*
gallons	2.31×10^{2}	cubic inches
gallons	3.785	liters
gallons of water	8.337	pounds of water
inches	2.54	centimeters*
inches	2.54×10^{-2}	meters*
inches of water	2.458×10^{-3}	atmospheres
inches of water	3.613×10^{-2}	pounds/sq. in.
kips/sq. ft.	4.882×10^{-3}	kg/sq. meter*
pounds	4.5359×10^{2}	grams*
pounds	4.448	Joules/meter* (Newtons)
pounds/cu. ft.	1.602×10^{-2}	grams/cu. centimeters
pounds/cu. ft.	1.602×10^{1}	Kgs./cu. meter*
pounds/sq. ft.	4.882	Kgs./sq. meter*
pounds/sq. ft.	4.788×10^{1}	Newton/sq. meter* (Pascal)
pounds/sq. ft.	4.788×10^{-2}	Kilonewton/sq. meter* (KPa)
pounds/sq. in.	6.895	Kilonewton/sq. meter* (KPa)
pounds/sq. in.	7.031×10^{-2}	kilograms/sq. centimeters
pounds/sq. in.	6.895×10^{3}	Newton/sq. meter*
pounds/sq. in.	2.307	feet of water
pounds/sq. in.	7.035	cm. of water
pounds/sq. in.	7.031×10^{2}	Kgs./sq. meter*
square feet	9.29×10^{2}	sq. centimeters*
square inches	6.452	sq. centimeters*
tons/sq. ft.	9.764×10^{3}	kg/sq. meter*

*SI Units (International System of Units)

CONVERSION FACTORS
METRIC TO IMPERIAL

To Convert	Multiply By	To Obtain
centigrade (degrees)	$(1.8) + 32$	farenheit (degrees)
centimeters	3.281×10^{-2}	feet
centimeters	3.937×10^{-1}	inches
centimeters of water	1.422×10^{-2}	pounds/sq. in.
grams	2.205×10^{-3}	pounds
cubic centimeters	6.102×10^{-2}	cubic inches
cubic meters*	3.531×10^{1}	cubic feet
joules/meter*	2.248×10^{-1}	pounds
kilograms*	2.205	pounds
kgs./cu. meter*	6.243×10^{-2}	pounds/cu. ft.
kgs./sq. meter*	3.281×10^{-3}	feet of water
kgs./sq. meter*	2.048×10^{-4}	kips/sq. ft.
kgs./sq. meter*	2.048×10^{-1}	pounds/sq. ft.
kgs./sq. meter*	1.422×10^{-2}	pounds/sq. in.
kgs./sq. meter*	1.024×10^{-4}	tons/sq. ft.
Kilonewton/sq. meter* (Kilopascal)	2.088×10^{-1}	pounds/sq. ft.
Kilonewton/sq. meter*	1.450×10^{-1}	pounds/sq. in.
liters	3.531×10^{-1}	cubic feet
millimeters	3.937×10^{-2}	inches
meters*	3.281	feet
newton/sq. meter*	2.088×10^{-2}	pounds/sq. ft.
pascal*	2.088×10^{-2}	pounds/sq. ft.
square centimeters	1.549×10^{-1}	square inches
square centimeters	1.076×10^{-3}	square feet

SUBJECT INDEX

AUTHOR INDEX